Nature in the Urban Landscape

The city, for all its congestion, for all its soot and smoke, is not a barrier that can stop the stream of life. It may slow down the movement of the stream, may even reduce it to a trickle but life remains and those of us with eyes to see can seek out and find its myriad branches almost wherever we chance to look.

John Rublowsky
Nature in the City
(Basic Books, 1967)

Nature in the Urban Landscape:
A Study of City Ecosystems

Co-authored by DON GILL *and* PENELOPE BONNETT

YORK PRESS / BALTIMORE

This book was manufactured in the United States of America.

Library of Congress Catalog Card Number 73-76409

ISBN 0-912752-03-3

Contents

Preface

Material for this book has been gathered over the decade 1962 through 1972, commencing when Dr. Gill began his graduate work at the University of California in Los Angeles, where he studied the urban incidence of the coyote.

Compilation of the bibliography and interviews with wildlife managers, ecologists, and conservation groups were conducted during 1970-71 by Penelope A. Bonnett for a graduate thesis at the University of Alberta under Dr. Gill's direction. A native of southeastern England, she brought to the book a first-hand familiarity with the London landscape and a knowledge of the English reports and projects cited in this book.

In 1971 the first Canadian study of urban wildlife was initiated in Edmonton under the auspices of the Canadian Wildlife Service, and is being directed by Dr. Gill.

Introduction

Studies of the behavior of people engaging in leisure activities remote from city environments indicate a reaction commonly approaching fear on the part of many individuals. The "average tourist" enjoying the National Parks rarely moves further than 100 yards from his car into the area adjacent to the highway. Because they have so little opportunity for direct contact with natural areas, many people have lost the art of knowing and appreciating nature. This seems likely to cost society dearly in terms of environmental damage. The present study suggests ways of re-establishing the links between man and his essential life-support systems.

The retention of natural landscapes within the city, together with their use for educational purposes, may enable many people who have lost their "rural roots" to enjoy and feel in harmony with nature—with those features and processes of the ecosystem which provide the ultimate constraints on any animal population.

In his large cities, man's activities have altered the physical and biological environment so greatly that he has created a totally new ecosystem. Increasing recognition is being given to this ecosystem, but rarely are the non-human organisms the focus of study; nature in the urban landscape is still an underrated resource whose esthetic and educational values have yet to be understood.

The city is primarily the habitat of man, but it is also utilized by a surprising array of other organisms, the majority of which go unnoticed by urban man. Little research has been conducted to investigate the interactions of man and his wild neighbors in the urban ecosystem. This type of research is necessary because urban areas may eventually function as important reservoirs for wildlife driven from its natural habitat (Gill 1965; Dagg 1970) and such wildlife will need proper biological management within the city.

The first book on urban ecology was written in 1945, when Fitter gathered together a history of the growth of London and its effects upon the indigenous flora and fauna. In a similar vein Kieran (1959) considered the natural history of New York City and described the amazing resilience of many species in the altered environment. Apparently few other cities have been subjected to such scrutiny and we therefore have little knowledge of the factors limiting natural populations in the man-made environment. We seek in this study to draw attention to the available information on urban flora and fauna with a view to assisting resource managers who may wish to enhance the natural elements in their cityscape.

The likelihood of an increase in the amount of green space incorporated in urban developments is considered in Chapter V. In the event that this increase is realized there will be an excellent opportunity for planners and wildlife managers to cooperate in the development of stable, natural communities as an integral part of the human habitat. This may be expected to markedly enhance the quality of the urban environment while at the same time contributing to the functioning of the regional ecosystem.

This study was motivated by a commitment to the idea that environmental quality can only be maintained or restored when a well-informed and ecologically oriented public applies pressure to its elected representatives (Brandwein 1966).

A considerable volume of literature in the past ten years has focused on the need to alter the methods of evaluating civilization's progress if mankind is to survive without further accelerating the rate of biospheric contamination (Commoner 1963; Boulding 1966; Caldwell 1966; Arvill 1967; Galbraith 1967; Dasmann 1968; Howard 1969; Platt 1969; Slatyer 1970; Spilhaus 1971). With the rise of affluence and the increasing amount of discretionary time available to urban dwellers there is a growing awareness of the need to maintain beauty in our everyday surroundings. Although quality of the environment has become a concern of many different groups in society, if the growth ethic is to be replaced by a desire to maintain a stable, self-renewing system, there is a need for each person to have the opportunity to enjoy direct communication with natural phenomena on an everyday basis. As Weinberg (1970) observed, we should eradicate the idea of preserving only National Parks because this has led us to the point where many of us live in urban squalor and enjoy beauty and nature only by travelling to these "holy shrines."

In view of the projected continuation of urbanization and urban growth it is vital to consider the implications of the ever widening gulf over which the average man interacts with nature. A fundamental question concerns the ability of those whose entire early experience is gained in the built environment to comprehend the intricacy of the ecosystem to which they belong. The quality of life in the future lies in the hands of an urban-based populace. It is the city dweller who will elect representatives who in turn will make the decisions that will drastically affect all natural resources.

The authors' contention is that environmental education, employing an ecological framework within which all studies can be taught, provides the main hope for a change in social values. If such an innovation is desirable, it is apparent that for many schools the resources with which to demonstrate ecological principles are not readily accessible. Greater attention should thus be paid to creating and enhancing "natural" habitats in population centers.

We believe that to enhance natural ecosystems within our urban environments is desirable. In this book we review significant facets of urban ecology and offer a relevant bibliography that we believe will be useful to those working toward this end.

Several questions were posed at the outset to clarify the basic issues involved in the concept of "planned nature":

1. Is it possible to maintain viable, natural communities in cities?
2. What values and conflicts result from the presence of nature in the city?
3. Do urban plants and animals possess special characteristics?
4. What are the principal factors limiting their existence?
5. Which of the existing management techniques are applicable to city vegetation and wildlife?
6. What are the implications of nature in the urban landscape to the planning profession?
7. What further research is necessary to establish the validity of the concept of increasing natural communities?

The chapters which follow provide some of the answers, but there is still a host of interesting studies to tempt the aspiring urban ecologist.

Nature in the Urban Landscape

CHAPTER ONE

The City Ecosystem

ECOSYSTEMS AND BIOTIC INTERACTION

A city is an ecosystem—an intricate web of interacting organisms involving energy transfer and materials cycling. In any ecosystem, a successful species is one that is able to channel energy efficiently and recycle materials so that an increasingly closed system develops that allows resources to be conserved and the species to be maintained. Species diversification stabilizes the system: the greater number of species, the more stable the system. The more complex the ecosystem (the greater the number of species), the less likely it is that any one species will grow explosively, and the more difficult it is for external forces to upset the ecosystem (Leigh 1965; Margalef 1968; and Klopfer 1969).

Man as a Perturbing Force in the Urban Ecosystem

In technologically advanced countries man has assumed, in large measure, the role of principal agent for change in the environment. He has diverted energy, water, and other materials from mature natural systems and used them to create his own unstable system. Man's activities which are important in modifying natural communities include removal of vegetation cover by tilling, logging, or burning; land drainage or inundation, earth moving and extractive processes; introduction of exotic species; pollution of air, water, and soil. Man-modified systems are distinguished by high energy inputs and turnover rates, lowered species diversity and low stability.

In cities you can see evidence of direct human effects upon plants—mutilated street trees, trampled and sometimes eroded open space, manicured ornamental gardens, a lack of both mature trees and leaf litter and a low species diversity. Animals are adversely affected by a great variety of our activities inimical to them, including illegal shooting, poisoning, trapping, damage to nests and young, and killing by high speed vehicles.

Of even greater significance are the general consequences of human activities such as housing and industrial development. These developments destroy established habitats, and habitat destruction is the greatest cause of diminishing animal numbers. Many plants with highly specialized requirements are excluded from populated regions. The removal of vegetation has far-reaching effects upon the animals which utilize it for an energy source. The general result is a great simplification of the ecosystem. By changing the characteristics of the environment, man alters the competitive relationships between species so that some are better able to survive in the new conditions than others. Natural communities in cities can therefore be expected to contain adaptable species typical of the early stages of ecosystem development.

Modification of the Physical Environment

Man has produced both readily visible effects and subtle and sometimes unperceived changes in the environment.

The emergence of large population centers with their inherent emphasis upon the artificial, built environment has meant a substantial adjustment in the natural communities which formerly occupied these areas.

Many species have been eliminated. Others have managed to exist in peripheral areas, only to be driven further away as habitat is destroyed with each new development. Perhaps of equal importance both to man and other creatures are the alterations which the city structure creates in the energy and water budgets in its vicinity. Associated with the growth of a city are certain processes of alteration:

1. Removal of plant cover and replacement by concrete, asphalt and other rock-like materials of greatly increased thermal capacity and reduced porosity.
2. Increased rates of surface water run-off, lowering of ground water reserves, and reduction in exposed water surfaces, marsh and other wetlands.
3. Dispersal into air and water of numerous inorganic and organic compounds and elements that modify the energy balance.
4. Release of heat energy as a result of combustion processes.

The effects of these processes are reflected in the distinct character of city climates. Many investigators have documented the measurable differences between the meteorological conditions of the city and its rural environs (Brooks 1952; Landsberg 1962; Chandler 1965; Geiger 1965; Lowry 1967; Maunder 1969).

The principal climatic changes in the typical city have been summarized by Maunder (1969) and are presented in Table 1. It is important, however, to recognize that within each urban climate there is a series of microclimates that result from local topographic and other controls such as the type of land use. For example, Chandler (1965) in his study of London demonstrated that open spaces in the

Table 1. Climatic Changes Produced by Cities

Climatic Feature	Comparison with Rural Environs
Contaminants	
dust particles	10 times more
sulphur dioxide	5 times more
carbon dioxide	10 times more
carbon monoxide	25 times more
Radiation	
total (horizontal surface)	15 to 20% less
ultraviolet (winter)	30% less
ultraviolet (summer)	5% less
Cloudiness	
clouds	5 to 10% more
fog (winter)	100% more
fog (summer)	30% more
Precipitation	
amounts	5 to 10% more
days with 0.2 in.	10% more
Temperature	
annual mean	1 to $1.5°F$ more
winter minima	2 to $3°F$ more
Relative Humidity	
annual mean	6% less
winter	2% less
summer	8% less
Wind	
annual mean speed	20 to 30% less
extreme gusts	10 to 20% less
calms	5 to 20% more

From Maunder (1969) after H. E. Landsberg (1960), *Physical Climatology.*

central zone displayed temperatures significantly lower than immediately adjacent streets; temperature differences between Hyde Park and nearby residential areas were as great as 2.3 degrees F. Such temperature gradients are at a minimum around noon and they range, over a twenty-four-hour period, from 0.5 degrees to 2.9 degrees F. in the London area. A transect from the outer suburbs through the central district and across the Thames to the southern periphery revealed the expected differences in temperature resulting from the high thermal capacity of the densely populated areas. Chandler (1970:88) in a recent essay about man's effect upon the climate, emphasized that

> A striking and certainly the most obvious departure from natural conditions is the canopy of solid, liquid and gaseous pollutants which covers many cities, particularly in winter . . . (and the) high concentration of carbon monoxide and other gases . . . from car exhausts which can exist for short periods in the bottom of deep urban chasms congested by slow moving traffic.

He also pointed out how airflow is altered by the presence of buildings and their relative positioning (Chandler 1970:90).

> The acceleration of winds along streets parallel to the free wind and the development of eddies across streets running at right angles to the wind are well known phenomena aggravated by grid-iron patterns of urban development and high-rise buildings.

These features of the city climate taken in conjunction with lowered humidity, reduced radiation, and increased frequency of light rainfall (Table 1) have important bearing on the survival of organisms living within the built environment.

Unfortunately, in today's urban ecosystems many of the material cycles remain unclosed, resulting in the accumulation of wastes which threaten the orderly existence of our civilization within the foreseeable future (Ayres and Kneese 1969). While much of the material dispersed into the air is carried by winds to areas remote from the source, a large quantity is returned to the city surface by rainfall and gravity. Thus we find acidified soil in urban areas, elevated levels of heavy metal elements and pesticide residue, and perhaps many as yet unperceived side effects of our tendency to ignore natural laws (Fahey and Butcher 1965; Thomas 1965; Purves 1966).

Effects of Changing Technology on Animals

Until the 1940s many organisms could adapt to the changing urban scene and maintain viable populations, but the rapidity and extensiveness of new technical developments since the 1940s has been too great for many species. A great decrease in city wildlife populations can be forecast if no action is taken to maintain their habitat.

The house sparrow (*Passer domesticus*) provides a good example of how a natural population can be influenced by technological change. The house sparrow is thought to have reached its peak population during the period of horse-drawn transport because of the abundant sources of food and nesting sites. The widespread use of wood in buildings and the intricate style of construction at that time resulted in a great variety of suitable nesting and roosting space. By the 1930s, with the replacement of the horse by the automobile, the sparrow was forced to seek new food sources and a noticeable decline in numbers occurred (Summers-Smith 1963).

The emergence of a clean and unfettered architectural style over the past thirty years has affected many birds which previously took advantage of the "cliff" habitat in the city. Bats and owls have also suffered from this change, as modern design rarely includes the kind of bell towers and spires that formerly provided roosting and nesting habitats.

Utility poles serve many birds as songposts, and provide lookouts for predators and potential sites for hole-nesting species. The wires strung between the poles and pylons are used extensively by fledgling birds and migrants; however, many birds meet their death when they cause short circuits between adjacent power lines, or fly into the lines (Harrison 1963; Benton and Dickinson 1966). Modern planning practice requires that utilities be placed underground in new communities and this will reduce essential habitat features for some birds. More birds may then turn to the metal lamp standard which has long been successfully utilized by the house sparrow.

The reduction in the use of wood as a building material affected not only the sparrow. Several other vertebrates and some insects were affected and excluded fron non-wood dwelling places. The house mouse *(Mus musculus)* is a very adaptable creature, however, and can be found wherever man settles, particularly in association with man's stored food. Cold-storage rooms have been colonized by mouse populations able to survive in freezing temperatures while exploiting the abundant source of fat and protein (Fitter 1945; Rublowsky 1967).

Some forms of advanced technology present problems which are not easily overcome by wildlife species. A fairly recent addition to the urban environment is the tall building or television mast extending far into the sky. Some reach heights greater than a quarter of a mile. Such structures possess warning lamps to minimize the danger to low-flying aircraft, but this unfortunately increases their danger to migrating birds which, attracted by the lights, crash into the building below the lights. Thousands of migrants may die at one location during each night of the short season when they follow traditional flight routes across continents. The erection of a tall building always causes a huge death toll until the birds learn to avoid the obstacle (Time 1970, 96[12]:10). In Wisconsin a typical example was the death of 5,595 birds in one night due to the presence of a new mast. In the worst instances as many as 15,000 casualties have been reported when migration activity is intense (Vosburgh 1966). Experiments with floodlights to illuminate the obstacle have shown that it is possible to reduce such deaths.

A famous example of adaptation to changing technology through natural selection is the case of the peppered moth (*Biston betularia*) in the industrial areas of Britain. During the nineteenth century in England intensively industrialized centers such as Manchester and Birmingham experienced loss of their lichen flora and the trees became coated with soot. The originally well-camouflaged pale variety of the moth now showed up clearly when resting on such dark tree trunks, and easily fell prey to insectivorous birds. The dark form thus held the competitive advantage and it has since become very common (Kettlewell 1958, 1961). This case of industrial melanism was the first in which natural selection had actually been observed in the field and emphasized the relatively short period over which such changes can occur. In the wake of recent successful air pollution control, pale forms of these moths are able to find surfaces on which they can rest and be camouflaged, and these pale forms are becoming more common.

The wealth of new materials now used by man has produced a change in the type of nesting materials available to birds. Species which frequent populated areas utilize a variety of objects including cellophane, paper, toffee wrappers, lemonade straws, steel wire cuttings, plastic-covered cable, and plastic sheets (Lond. Bird Rep. 1963, 26:73). New materials may also be used for purposes other than nest construction. Hawkins (1970) reported seeing a brooding moor-hen (*Gallinula chloropus*) using a piece of plastic sheet as a protection against rain showers!

The scarcity of open water in cities has caused widespread extirpation of reptiles and amphibians from urban sites. The drainage of pools during development results in

Islands in artificially created ponds or reservoirs can provide urban nesting sites. Here a Canada goose (*left*) nests with a great crested grebe on such an island, with Mallard ducks in the background. (*Photo by Pamela Harrison, A.P.R.S.*)

loss of breeding habitat for common frogs (*Rana temporaria*) and toads (*Bufo bufo*). Woollacott (1961-67) believes that the reduction in habitat causes a great increase in competition between species (a common form of which is egg eating), and this further explains the scarcity of these animals in urban areas.

Other organisms have shown themselves more adaptable. The numerous sources of heated water in cities have proved an attractive habitat for many aquatic species. In London it is reported that even some species of tropical fish are found breeding near power station outfalls (Lond. Nat. 1969, 48:85).

Plants and Animals as Health Hazards

The air contains numerous viable particles which may cause disease or allergic reaction in susceptible people. Fur and feather fragments are allergens to a few people, and ragweed (*Ambrosia* spp.) pollen is known to cause misery to countless individuals each year. Bacteria, fungal spores, and insects also may have detrimental effects upon humans. It is interesting that urban green spaces, although acting as potential sources of pollen and fungal spores, are measurably less contaminated with bacteria. In New York City bacteria counts in the air showed from 3 to 30 bacteria per cubic foot, and the lowest values were found in Central Park where on average the number was only one-third that found in street air (Jacobson 1968).

Many birds and animals are potential carriers of disease and therefore could pose a threat to public health. Virulent strains of fungal infections have been detected in dried and weathered pigeon manure taken from old nesting and roosting sites, as well as from fresh droppings and soil in a Washington, D.C., park. Starling (*Sturnus vulgaris*) roosts are suspected of being reservoirs of fungal spores capable of causing pulmonary infection (Jacobson 1968). Some workers are of the opinion that feral pigeons (*Columba livia*) living in cities represent a distinct health hazard because they can provide a reservoir of ornithosis virus. However, little direct evidence exists that humans have contracted disease from such sources (Lepine and Sautter 1951; Hughes 1957; Meyer 1959; Thearle 1968).

All warm-blooded animals are potential rabies carriers. Animals which might present such a hazard in cities include not only dogs and cats, but also bats (*Chiroptera*), coyotes (*Canis latrans*), foxes (*Vulpes* spp.), opossums (*Didelphis marsupialis*) and skunks (*Spilogale putorius* and *Mephitis* spp.).

Property Damage by Plants and Animals

Another danger inherent in maintaining natural areas in cities is the increased likelihood of aircraft loss due to birdstrikes. If birds congregate in and around airports they can be a hazard to planes; however, there have been only two instances in the United States where this has caused loss of civilian lives. Costly damage to planes has been widespread, however, particularly since the advent of jet engines. Management techniques have been devised to minimize the attractiveness of runways as feeding and roosting habitat. Other forms of control have aimed at removing the birds by frightening them away prior to the runways' being used (Munro and Harris 1963; Drury 1966; Solman 1966; Wright 1967; Czaja 1968; Murton and Wright 1968).

A less fatal but more common sort of property damage results from birds using buildings for nesting and roosting which may cause esthetic and mechanical damage from deposits of excreta on masonry and pavements. Utility companies are believed to suffer considerable losses from the activities of birds which excavate nest holes in posts and cause breakdowns through the creation of short circuits between power-carrying lines (Turcek 1960; Vosburgh 1966). Rumsey (1970) has recorded the preference of the red-headed (*Melanerpes erythrocephalus*) and pileated (*Dryocopus pileatus*) woodpeckers for freshly creosoted poles in which to nest.

Mammalian species can damage growing plants in gardens and parks. The gray squirrel (*Sciurus carolinensis*), for example, is prone to strip the bark from trees at certain times of the year (Shorten 1962). Suburban foxes (*Vulpes vulpes*) in England, because of their fondness for playing in sand bunkers, are unpopular with maintenance staff at golf courses. These animals have been reported to carry away golf balls much to the annoyance of players (Vesey-Fitzgerald 1965; Teagle 1967). Foxes also sometimes inspect and upset garbage containers in suburban areas.

Plants too, may damage buildings when the growth habit of a given species is not understood by the property owner. On clay soils roots of willow (*Salix* spp.) or poplar (*Populus* spp.) close to buildings can withdraw water to the extent that soil shrinkage results in damage to foundations (Min. H. L. G. 1958; Salter 1960; Zion 1968; Jaffa 1970). Herbaceous species can force their way through asphalt and concrete surfaces on roads, tennis courts, and paved walks. Elm trees (*Ulmus* spp.) are well known for their lack of stability once they become mature, and during gales they may be whipped against buildings.

Simpler plants such as fungi or the unique organisms known as slime-molds can live on damp wood and cause it to decay; thus they hasten the demise of old buildings that have damp surfaces (Ing 1964).

THE INFLUENCE OF THE
URBAN ENVIRONMENT

Effects of the Urban Physical Environment on Plants

To grow and reproduce, a green plant requires light, carbon dioxide, and water, in addition to a variety of elements and compounds contained in the soil. The layer of the atmosphere in which most vegetation grows (the first five feet above the surface) differs greatly from the air mass which extends above. Within this lower layer there is environmental differentiation of surprising magnitude (Geiger 1965). To a large extent the nature of the plant cover determines microclimatic conditions (within basic topographic controls), and it is therefore possible to have a variety of microclimates within short distances if the vegetation cover forms a heterogeneous mosaic.

Of the several climatic conditions altered by the city environment, the one with the greatest potential importance for plant growth is that of reduced light received by plant surfaces. There is little information on this topic and, considering that light is the basic energy source for the food chain, it is important that studies of urban plant productivity be initiated. It can be presumed that the rate of photosynthesis will be reduced in a city environment, but the magnitude will depend upon several other factors.

Reduction in light results primarily from the dust haze which usually hangs above a city. Part of the dust in the air is deposited upon foilage, and this further reduces the radiation received by the photosynthetic tissues. Annual dustfall in cities is around one-fourth pound per square yard and can reach two pounds per square yard in heavily industrialized areas (Mellanbv 1967).

An interesting question, and one which is worthy of more study, concerns the possible effects of artificial lighting on plants growing in streets and gardens. An artificially lengthened day, even if the light is of low intensity, influences the flowering of many plants. Light from a street lamp may cause a tree to burst into bloom even when the natural days are too short to cause flowering. Photosynthesis, on the other hand, is less likely to be influenced by artificial light, since most plants require rather intense light for that process.

One consequence of the elevated temperatures found in the urban environment is the lengthening of the growing season for plants; the potential increase can be as great as five weeks in some city areas (Landsberg 1956). However, an actively growing plant is likely to be more affected by pollution than a quiescent one, and thus plants may suffer proportionately more damage at the limits of the growing season in cities. This is one factor contributing to the observed susceptibility of many evergreen species to atmospheric contaminants in towns.

Reduced humidity is most likely to affect the simple plants, e.g., lichens and mosses, whose ability to resist desiccation is limited by their surficial habit. Lichens are commonly found in "pioneer" situations colonizing rocky surfaces in most regions of the world. This ability to survive in barren habitats comes from possession of mechanisms which allow them to efficiently absorb and concentrate nutrients.

Humidity conditions in autumn and spring are conducive to lichen growth. Unfortunately in the city this is also a time when air pollution is high. High sulphur dioxide concentration in particular appears to affect the distribution and viability of lichens. Microclimatic measurements have revealed that the average moisture content of the air diminishes rapidly with height above the surface (Geiger 1965) and Skye (1968), working in Stockholm, noted the restriction of many lichen specimens to the lower portions of tree trunks as the central city was approached. It thus appears that for simple plants lowered humidity may be important, but its effects are commonly masked by the more drastic influence of pollutants in the urban atmosphere.

One factor contributing to lowered moisture in the city is the paucity of evaporating surfaces due to the lack of vegetation. During certain periods, many urban trees are undoubtedly subject to considerable stress because of the rapid loss of moisture induced by turbulence, low humidity, and high temperatures, in addition to low soil-moisture levels.

Interdependence of the factors that influence growth is further underlined when the effects of increased amount and frequency of rainfall in urban areas are considered. Precipitation washes the atmosphere and carries down soluble gases and dust. This has produced acidity in urban soils. At times acidic rain may damage plant tissues in severely polluted areas (Thomas 1965). Soil acidity influences the species of earthworms living in the ground; few species are found where the pH is lower than 4.0 (Satchell 1967). Acidification of soils has the effect of making certain key nutrients (calcium, ammonia, and nitrates) less available than they otherwise would be; in addition, the acid condition restricts microbial and enzymic activity, thus leading to further nutrient deprivation of the soil.

The reported increased frequency of light showers in cities could have important implications for plant growth.

It could result in shallow rooting due to lack of water penetration and rapid evaporation of surface water. In addition, if the amount of rainfall at any one time is insufficient to wash away the accumulations of dust and other contaminants and merely adds further acidic materials, then foliage may benefit little from precipitation. The underside of foilage is less completely washed than the upper surface, and may suffer excessive build-up of contaminants.

In summary, it appears that few studies of plant ecology and physiology have been carried out in urban areas. We can nevertheless assume that no single factor will limit productivity, but, dependent upon season, type and degree of air pollution, and soil moisture and nutrient availability, reduced sunlight reaching the ground should result in decreased photosynthesis.

Effects of the Urban Physical Environment on Animals

Since animals are dependent upon green plants for both energy and cover in their living and reproductive activities, the direct effects of the modified physical environment are even more difficult to perceive than in the case of plants.

Little evidence exists about the effects of reduced radiation loads on urban animals, but it seems unlikely that this is of major physiologic significance. The extension of "daylight" through the use of artificial lighting, however, does appear to have an influence upon songbird activities. It is reported that English blackbirds (*Turdus merula*) sing at night in central London and have been found nesting during the winter in Berlin (Grummt 1962; Mitchell 1967; Blackett 1970). Both of these occurrences were attributed to the effects of street and display lighting. Presumably extended "daylight" can provide the requisite stimulus to initiate physiological changes culminating in earlier gonadal activity. In support of this is European evidence that suburban blackbirds breed at least ten to fourteen days earlier than their rural relatives (Snow 1958; Havlin 1963). There have been several reports of blackbirds in central London singing as early as January and as late as August, which suggests a greatly extended breeding season for these urban individuals (Lack 1944; Lloyd 1944; MacAllister 1944; Mitchell 1967). Wood pigeons (*Columba palumbus*) may nest four weeks earlier in city locations than they do in woodland (Cramp 1968), and robins (*Erithacus rubecula*) show a similar tendency (Lack 1965). It is probable that the effects on songbirds mentioned above are also in part a function of increased temperatures in the city environments. The greatest effect of increased urban heat occurs in the winter, when the minimum temperatures are significantly higher in the city than in rural areas. Birds particularly benefit from the enhanced temperature which reduces winter mortality brought about by lack of food and by cold stress (Lond. Bird Rep. 1963, 27:5; Snow 1967). The sight of city pigeons huddled over a street grating on a chilly morning is evidence of the birds' need for warmth and their adaptability to the urban environment. There is little documented information on this topic but it may be assumed that waterfowl can use city waters as feeding and roosting areas due to the presence of open water (Hochbaum 1965). For example, in Rochester, Minnesota, warm water discharged by a local power plant keeps a lake from freezing. Giant Canada geese and other waterfowl can be seen there during the winter months.

A low humidity does not ordinarily have serious consequences for mammals but it does influence the distribution of many invertebrates that require moist conditions for survival. Creatures such as earthworms, slugs, centipedes, wood lice, and so on, may have a very limited choice of living sites in the inner city, but the suburban mosaic of backyards, gardens, parks, and school grounds provides a multitude of suitable habitats (Barnes and Weil 1944,

1945). Many insects have early life stages requiring water; in densely populated areas they may find such a commodity in scarce supply.

Reduced plant cover, together with lower humidity, makes life difficult for birds that exist by excavating soil fauna. Snow (1958) reported heavy nestling mortality during June in England among blackbirds, due to the dryness and compaction of lawns and flower beds in which earthworms reside.

Pollution and the Urban Terrestrial Ecosystem

Man's activities pollute his surroundings and the effects are widespread and important both for man and the other organisms living in his built environment. It is not always possible, however, to distinguish the effects of the changed physical environment of the city from those caused by ambient pollution.

A city which has evolved over several centuries contains within its boundaries a mixture of land uses, with industry—usually a major source of pollution—located throughout the urban fabric, although it is particularly aligned along transportation routes. Residential property is the other main emission source. While it is clear that stationary sources are ubiquitous throughout the built-up area, mobile sources of pollution are equally important. Some 50 to 60 percent of all air pollution stems from such sources as gasoline-powered automobiles which have access to most of the built environment (Ayres and Kneese 1969). New emission standards such as those being imposed on automobiles in the United States may change the relative significance of pollution from mobile sources.

Not all cities are equally affected by the various forms of pollution, since the intensity of effect is a function of geographical location, population size, types of fuel burned, and nature of industry, in addition to local topography and climate. For example, the air pollution problems experienced in Los Angeles are partially the consequence of the high intensity of light radiation there and the location of the city on the coastline hemmed in by mountains. Photochemical oxidation of emitted substances produces secondary products that have a toxicity considerably higher than that of the original compounds. For some indication of the automobile as a source of pollutants, reference should be made to Stern (1968, Vol. III, pp. 55–95).

Plants could be used as indicators of general contamination in urban (and other) areas. This might allow trends of increasing pollution to be detected at relatively low cost so that control may be initiated before public health is affected.

Plant growth is inhibited by air pollutant concentrations far below those at which obvious signs of damage occur. Photosynthesis is inhibited to varying degrees by many gases, including sulphur dioxide, ozone, hydrogen cyanide, and hydrogen fluoride. In zones of heavy particulate contamination, photosynthesis may be reduced to one-tenth the normal rate due to coating of the foliage by dust (Mellanby 1967). Certain plants (e.g. citrus) are very susceptible to smog, yet others (e.g. gladiolus) produce resistant varieties capable of withstanding greatly elevated pollution levels (Rich 1964; Darley 1969).

One of the photochemical secondary products derived from automobile emissions is ozone, the normal background concentration of which in the lower atmosphere is only two to three parts per hundred million (pphm). Experiments with tobacco and pine have revealed damage to tissue when exposed to a level of 6 pphm for six hours (Katz 1967). Measurements in urban areas indicate that ozone levels of 32 pphm can occur in the vicinity of congested streets (Tebbens 1968).

Other pollutants known to produce damage to foliage,

inhibit photosynthesis, alter enzyme activity, and cause leaves to fall, are peroxyacetyl nitrates (PAN), ethylene, nitrogen oxides, sulphur oxides, fluorides, and particulates. Not all of these derive from the automobile, but all are commonly to be found in urban atmospheres (Brandt and Heck 1968).

As was noted previously, plants growing in the city may experience a longer growing season, and in an area where pollution is present this may be disadvantageous. Thus conifers (*Gymnospermae*), for example, are much more susceptible to injury and death in cities due to their evergreen habit. According to Darley (1966), the principal effect of particulate contamination is the clogging of microscopic leaf openings; this reduces both photosynthesis and respiration, and eventually leads to stunting and death. Those evergreens that hold their leaves for six to eight years suffer most in a heavily polluted location (Fitter 1945; Skye 1968). Thus pine (*Pinus* spp.) withstand city life better than do spruce (*Picea* spp.) because pine trees retain their leaves only half as long as spruce. The conifer most tolerant to pollution is the larch (*Larix* spp.), which is deciduous.

The ability of some species or varieties to survive in regularly polluted areas usually results from a metabolic mechanism that renders substances innocuous once they enter the leaf tissue. Thus, in Sweden, Lihnell (1969) found that sulphur dioxide is converted to sulphate compounds by birch (*Betula* spp.) and apple (*Malus* spp.) trees. Leaf analysis revealed concentrations of sulphur some two to nineteen times as great as would be found in specimens from an unpolluted area. Some deciduous plants are able to store excess compounds in a harmless form and shed them with their leaves in the autumn, thus giving them a further advantage over evergreens.

Observations of trees in the major squares of Paris exemplify the importance of air contaminants in controlling vascular plant reproduction (Tendron 1964; Soulier 1968). It was recorded that about 1927, before World War II, the blooming of the princess trees (*Paulownia* spp.) became abnormal due to buds dropping off before they opened. During the war when both industrial and domestic combustion sources were reduced and private cars largely immobolized due to lack of fuel, the trees regained their former springtime splendor and subsequently retained their fruits to the normal season. However, shortly after the end of the occupation when economic activity returned to normal and atmospheric contamination levels rose again, the trees once more suffered bud-fall and erratic blossoming periods. The principal components believed to cause these changes are carbon monoxide and ethylene which act together to influence plant growth hormones.

One of the most studied groups of organisms with respect to air pollution are the lichens (Skye 1958; Brightman 1959; Fenton 1960, 1964; LeBlanc 1961; Gilbert 1965; Pearson and Skye 1965). They are susceptible to damage from contaminants in the air and in rainwater. The scarcity of lichens in cities is highly correlated with the presence of industry and high-density populated areas. It is possible to show close correlation between the absence of certain lichen species and the ambient levels of sulphur dioxide (Gilbert 1965, 1969, 1970). Lichens can be used as valuable indicators of air pollution (Gilbert 1970; see also Lihnell 1969).

Skye (1968) noted that the pH of tree bark is often sufficiently acid to preclude the presence of lichen vegetation in central Stockholm. The rain tracks on city trees are usually devoid of lichens, whereas horizontal branches may possess quite rich lichen cover due to the more favorable substrate they offer.

The principal effect of sulphur dioxide upon lichens is thought to be photosynthesis-reducing damage to the algal chloroplast. Mosses are similarly affected; Syratt and Wanstall (1969) claim that sulphur dioxide affects certain mosses more than others, the difference stemming from varied abilities to metabolize the gas to sulphate compounds. Species that can withstand high levels of sulphur dioxide possess large amounts of chlorophyll and in addition are able to convert it to sulphates in large quantities without harm to the tissues.

Other simple organisms may also be excluded from urban areas due to high pollution levels. The effects of pollution on urban fungi are poorly documented, but it appears that basidiomycetes are not well represented in cities; this may stem from low humidity and unsuitable soil conditions. The lack of dead timber in the city green spaces is also important in this respect. Molds and rusts, while not commonly seen on city vegetation, are often observed in homes, indicating the presence of viable spores in the atmosphere. That fungal infection is restricted by urban environments is demonstrated by the rarity in cities of two plant diseases that are common in non-urban areas—needle cast (caused by *Lophodermium juniperinum*) on juniper (*Juniperus communis*) growing in polluted districts of London, and black-spot fungus (*Diplocarpon rosae*) on roses growing in polluted areas in industrial sites (Air Pollution 1969, p. 238).

Lead in paint poses serious danger to children and is a well-known public health problem. Lead is also a commonly occurring air pollutant which, above certain critical levels, no doubt can cause illness in man and other animals, and can adversely affect plants. The spur to interest in lead as an air pollutant came from the discovery of elevated levels of this element in soils and vegetation along heavily used highways. Isotopic identification has revealed the source of the lead as lead alkyl additives used in gasoline and emitted in to the atmosphere in exhaust fumes (Cannon and Bowles 1962; Purves 1966, 1967; Ruhling and Tyler 1968; Kloke and Leh 1969; Chow 1970; Lagerwerff and

Specht 1970). Accumulations of other metals such as zinc, mercury, cobalt, and nickel are now being studied.

There is a scarcity of information about the effects of the increasing dissemination of heavy metals occurring today both in cities and in rural areas. An interesting study conducted in Pennsylvania may foreshadow an increased use of wildlife in cities as pollution indicators. Tansy and Roth (1970) recorded the levels of metallic lead in the hard and soft tissues of city pigeons and compared these to levels found in rural populations. They showed that pigeons from central Philadelphia had two to five times the amount of lead found in the tissues of rural specimens.

Mosses have been used as indicators of lead levels in the environment because they rely upon precipitation (rain, etc.) for most of their mineral nutrients. Work in Sweden has shown that roadside specimens contain lead in concentrations of 300 to 500 ppm. Following a study of the lead content of mosses growing in central and southern Scandinavia, Ruhling and Tyler (1968) concluded that lead concentration in mosses increases with precipitation and with decreasing distance to large population centers. By analyzing museum specimens and material from peat bogs, the Swedish researchers concluded that the background levels of lead are small, and the currently measured concentrations in urban and industralized areas are the result of human activities.

Brandt and Heck (1968) and Otto and Daines (1969) have illustrated how plants are affected by pollutants in varying degrees depending upon other environmental factors existing at the time of exposure. Thus, soil nutrients, soil moisture, and relative humidity, in addition to light intensity, will all modify the effect of a pollutant. It was shown, for example, that injury is reduced in plants sensitive to a pollutant if they are relatively dry during exposure. The generally dryer air to be found in the city may thus counteract potentially damaging pollutant levels to some extent.

From these scattered observations of the interaction of vegetation and air pollutants it may be concluded that where traffic congestion is a regular feature and both domestic and industrial emissions concentrate, plants are likely to suffer reduced growth and may be unable to reproduce.

Considering the effect of environmental pollution on urban wildlife, it should be recalled that all animals are ultimately dependent upon plants for food, and most animals require vegetation for cover and nesting. Any change which brings about a modification in the plant species present and in the abundance of plants may then have an effect on animals higher up the food chain, whether they are herbivores, carnivores, or omnivores. It is therefore possible for indirect as well as direct effects to be experienced by animals living in a polluted location.

Much of the experimentation in connection with pollutant effects has been done with laboratory animals such as rats, mice, guinea pigs, rabbits, and chickens. The emphasis has been on the determination of toxic levels and it is clearly unreasonable to extrapolate directly from such studies to the wildlife which exists in cities. Nevertheless, it is probable that chronic effects due to air pollution are present in many urban animals and have yet to be investigated.

Most of the substances that are known to be present in urban air have been subjected to laboratory studies. From these experiments, we know that ozone, a toxic, fast-acting pollutant resulting from photochemical oxidation of automobile emissions, has long-term potentially insidious effects on all forms of animal life. Tests with mice and rabbits (Stokinger and Coffin 1968) showed that low-level daily exposure to ozone, or infrequent high-level exposure, produced permanent lung damage. In addition, the mortality rate was increased in litters where the mother had suffered such exposure during pregnancy.

Sulphur dioxide in the urban atmosphere is unlikely to cause permanent detrimental effects at its present level of concentration (Stokinger and Coffin 1968). However, reports from Czechoslovakia from an industrial area where this gas has periodically high concentrations indicate that hares (*Lepus* spp.) may suffer lowered reproduction rates (Nováková 1969).

Nitrogen oxides are not considered to reach levels dangerous to the public health in cities at present. Carbon monoxide, however, a component of exhaust fumes from many combustion processes, is highly toxic. Current research suggests that low-level cumulative effects occur only rarely in either humans or laboratory animals; but whether or not this holds true for urban species such as sparrows and pigeons, which frequent roadsides where they may be exposed to high levels of carbon monoxide, has not yet been investigated. There is evidence that adaptation to low-level exposure does occur in dogs; it may be possible that truly urban animals will develop a tolerance to carbon monoxide in the air (Morrow 1964).

There are now documented cases of both fatal and chronic effects on animals exposed to severe urban air pollution. Two of the most infamous occurrences took place in Donora, Pennsylvania, in 1948, and in London, England, in 1952 (Stokinger and Coffin 1968). In addition to recording large numbers of human fatalities some documentation of effects upon pets and livestock was also made. It was noted in both cities that canaries and dogs were highly susceptible. Chickens and cats seemed less affected but many cows were reported to have suffered acute respiratory distress and a few died of heart failure.

Veterinary research comparing rural and urban pets has produced some interesting results suggesting that dogs suffer from diseases and ailments attributable to their polluted environment. In a three-year survey of 1,007 dogs in Philadelphia and its suburbs, Reif and Cohen (1970) showed a higher frequency of pulmonary disease in city dogs than in rural ones. This was most noticeable for middle-aged and older animals. Ragland and Gorham (1967) demonstrated the prevalence of carcinoma of the tonsil in city dogs and suggested the need for studies to establish the relationships of atmospheric carcinogens to the distribution of affected animals.

Pollutants may adversely affect the sensory receptors of animals. Mellanby (1967) points out, for example, that urban foxes may be suffering a detrimental decline in the acuity of their sensory organs.

The high noise levels in cities may be responsible for stress and for the deterioration of hearing in human beings. Sonic booms (which, of course, are not limited to urban areas) do harm animals. Abortions in sheep and mink and reduction of lactation in cattle have been reported, for example. But we were unable to find reports of any work on the effects of ordinary urban noises on animals. From records of nesting and breeding sites there is little indication that animals or birds shun noisy situations. Foxes are recorded raising litters under a shed on a building site and on railway embankments; songbirds have nested in trucks, aircraft, cranes, tractors, and cars, even when these are still in active use (Bridgman 1962; Brown 1963; Gladwin 1963; Teagle 1967, 1969). Land adjacent to heavy industrial premises or to airports where there is attractive plant cover is utilized by a variety of creatures which appear unperturbed by high noise levels (Murton and Wright 1968).

With respect to use of pesticides in the urban environment, it has been suggested that less use is made of insecticides, herbicides, and poisons in cities than in intensively farmed regions. Although no documented evidence for this has been found, intuitively it appears reasonable.

However, affluence of the suburbs and obsession with cleanliness may result in locally heavy pesticide use (Niering, 1969). Tarrant and Tatton (1968) studied levels of pesticide residues in samples of rainwater taken at widely separated locations in Britain over a twelve-month period. They showed that urban atmospheres had relatively great amounts of chlorinated hydrocarbons absorbed onto the large number of suspended carbon particles present. London rainwater was more contaminated than that collected at rural sites where extensive agriculture was practised, but somewhat lower than in districts such as the Kent fruit-growing area.

Many municipalities have spray control programs for weeds and fungal diseases, as well as for mosquitoes. Tabor (1965, 1966) reported that pesticide levels of urban atmospheres are sufficiently high in some parts of the United States to represent a potential public health hazard.

The residues and metabolites of DDT are less directly toxic than their precursor but are equally long-lived and capable of physiological effects that perturb ecosystems (Anderson *et al.* 1969). Many raptors, particularly those preying upon other birds or fish, have accumulated concentrations of DDT and its metabolites to levels that have rendered some species infertile due to disturbances in the hormonal system controlling egg production. With the advantage of hindsight it is possible to suggest that the first symptoms of DDT accumulations were recorded in 1949 by Hall, who studied peregrine falcons (*Falco peregrinus*) nesting on a building in Montreal (Hall 1970). These birds had successfully nested at this site since 1936, but suffered increasing difficulties from 1949 to 1952, when they finally disappeared from the city locale. Use of this biocide in Wisconsin cities to control Dutch elm disease resulted in 80-90 percent mortality of local American robin (*Turdus migratorius*) populations that fed upon earthworms in the soil beneath the sprayed trees (Hickey and Hunt 1960;

Wallace *et al.* 1961; Hickey 1966). Predatory invertebrates may similarly accumulate toxic levels of these compounds, and a depletion of their numbers is usually followed by an increase of their prey population, which may then become a pest (Edwards 1965).

Accumulated chlorinated hydrocarbons are likely to become especially important for animals in the wild during the winter and spring, when food is scarce and temperatures are low. Utilization of fat deposits under these conditions may release large quantities of the toxin into the blood stream, and death can result. Recently, Cooke (1970) documented experiments in which tadpoles of the common frog were shown to be most susceptible to stored DDT at the time of metamorphosis.

There is also an industrial and domestic source of chlorinated hydrocarbons. Polychlorinated biphenyls (PCB) find their way into natural systems by emissions and effluents from urban areas where they are used or manufactured and thus may represent a particular hazard to animals frequenting these places. The close chemical similarity of PCB to DDT makes their potential effects somewhat predictable, although experimental work indicates that effects such as shell-thinning (a common result of DDT ingestion) do not result from high levels of PCB in tissues of birds (Risebrough *et al.* 1970). Nevertheless, delayed breeding in birds may be attributable to these industrial biocides.

Clearly we need a great deal more information about levels and effects of chlorinated hydrocarbons in urban ecosystems. For many birds and small mammals only the more immediate toxic effects of pollutants are relevant, since these animals are relatively short-lived and may not experience cumulative effects. The average songbird lives only one to two years and thus is unlikely to display signs of chronic pesticide poisoning. Few wild creatures enjoy their potential life span since it is usually cut short by accident, disease, or predation. However, the long-term effects may be significant for larger mammals whose average life span is in the order of ten years.

Pollution and Urban Aquatic Ecosystems

Historically, urban development has concentrated near water bodies which provide drinking water supplies, transport, power, and a diluting medium for wastes. Since in many instances the accompanying flat land was subject to periodic flooding, it was expedient to dike and drain extensively to facilitate building. In the name of progress, coastal areas have suffered considerable modification by filling to create more land.

This ruined structure illustrates the danger of building on low-lying land that is subject to floods. Built next to a stream near Baltimore, Maryland, this thriving commercial enterprise was devastated by a flood in 1972. Usually it is advisable to keep flood plains as natural areas. They often can provide excellent linear habitat for wildlife and can contribute to the ecological stability of a locality. (*Silvan photo.*)

The outcome of years of development is a dearth of wetland and open water in many urban regions, except perhaps for a number of small ornamental pools. In addition, extensive sections of rivers have evolved as open sewers incapable of supporting a living community save for the lowliest members of the decomposer chain. Introduction of large amounts of organic material into waterways has resulted in depleted oxygen levels due to excessive bacterial decomposition. In less drastically changed waters there is a trend toward greater productivity due to the increased input of heat energy and nutrients in the form of effluent and sediments. The total effects of urban growth upon water resources vary considerably, depending upon the degree and type of industrialization associated with the settlement.

In this century we have witnessed a growth in the rate at which metals and metal compounds are dispensed directly into water courses and indirectly via rainfall. Persistent biocides now contaminate all waters. The common result of the several inputs to the aquatic system is simplification of the community of living organisms. Species typical of the early successional phases are abundant and productivity is high (Hynes 1960, 1965; Mackenthun 1969; Wilbur 1969).

Construction in urban areas often gives rise to erosion, producing large quantities of sediments in streams and rivers of the watershed. Wolman and Schick (1967) calculated that developed areas in Maryland produced from 1 to 75 times more sediment than rural agricultural districts. Similar figures are available for other East Coast locations (Felton and Lull 1963; Guy and Ferguson 1970).

Often unrecognized sources of water pollution in residential areas are surface drains which handle large quantities of oil and grease from vehicles, de-icing compounds applied to roads, road marking paint containing metals such as chromium, detergent from on-street car washing, and the excreta of pet animals.

Dogs in New York City number about 500,000 (with about 282,000 registered), while London boasts a total of about 700,000 registered dogs (Time 1970, 96[3]:35). With several hundred thousand dogs living in a city such as New York or London, we estimate that the minimum quantity of urine finding its way into the city environment each day is 22,000 gallons (New York) and 30,800 gallons (London). Similarly the amount of feces deposited would be in the order of a minimum of 125,000 pounds per day in New York and 175,000 pounds per day in London. Clearly this makes a significant contribution to the amount of untreated sewage reaching city watercourses and may considerably influence the metabolism of urban vegetation. (See also Beck 1973.)

Interaction of Organisms in a Polluted Environment

The pronounced influence of chemical pollutants upon vegetation must be important to animals utilizing it. Stokinger and Coffin (1968) provide data on the reduced health of livestock which had been fed contaminated forage. In areas of fluoride contamination, with ambient fluoride concentrations of parts per billion (ppb), it is possible for vegetation to accumulate toxic levels over a long period of time.

Deer (*Odocoileus virginianus*) in Canada have been reported suffering poor health due to the proximity of an industrial plant emitting fluorides. The animals were seen to drink regularly from effluent ponds containing fluoride concentrations of 4000–7000 ppm. Few animals lived beyond two years of age and many had broken teeth and fractured jawbones (Karstad 1967). In Czechoslovakia livestock showed high fluorine levels in the bones, teeth and milk when the animals were exposed to polluted air. Sparrows, frogs, and rats in the area were found to contain two to fourteen times the concentration of fluorides detected in similar animals from a clean area (Balázová and Hlucháň 1969). Other substances such as arsenic, lead, molybdenum, beryllium, manganese, zinc, and selenium can concentrate in vegetation and become potentially hazardous to animals. From Germany come reports of deer and hares poisoned by eating arsenic-contaminated vegetation near smelters (Air Pollution 1969). In general, however, due to various protective physiological mechanisms, there tend to be fewer ill effects from ingestion of contaminated vegetation than from direct inhalation of pollutants.

The species diversity of insects is markedly reduced in districts with heavy pollution. Vegetation coated with particulates will support less animal life than clean plants, and polluted water may not provide suitable conditions for those insect species that have aquatic larvae. Lack of specific food plants curtails the number of moths and butterflies able to live and reproduce in such an area. For example, Nováková (1969) has produced preliminary evidence for a decreased number of insects in industrialized areas of Czechoslovakia.

Reports from London suggest that particulate pollution is an important factor limiting flying insect populations (Gooders 1968). Since the introduction of smoke controls in 1954 there has been a noticeable increase in the number of insect-eating swifts (*Apus apus*) and house martins (*Delichon urbica*) in districts close to the city center. In the past these birds were seen to be gradually retreating from the inner city as a result of the presumed lack of food. Gooders (1968), however, believes that the return of these birds is directly related to reduced particulates in the air. In view of the swift's mode of feeding, which involves flying open-mouthed and collecting all objects in the flight path, it is not surprising to find it absent from smoky areas. Soot particles, heavily contaminated with sulphur compounds, may be devastating to a bird such as the swift. Improvements in air quality may encourage an increase in the flying insect population, which would also benefit the swift.

The current appraisal of herbicide use suggests that it has rarely had direct effects upon animal life; but certainly it has been instrumental in removing particular plant species from local flora, thus destroying necessary food or cover for animals (Moore 1965; Way 1969). An exception to this is the influence of chemical spraying on bees (*Apis* spp.) whose essential pollinating activities have been halted over extensive areas through mortality. Use of sprays during the flowering period means that toxic concentrations can collect in the base of the flower cup; then when bees take nectar they also absorb high concentrations of chemicals (Way 1969). Such mortality can be avoided by careful timing of spraying to avoid the flowering period. By a similar mechanism bees are exposed to fatal doses of arsenic, zinc, and fluorides in areas where these metals occur in aerosol form (Tendron 1964).

One detrimental result of herbicide use observed in livestock, but potentially of interest when considering wild herbivores, is the increase in palatability exhibited by usually toxic plant species. Through some unknown mechanism, herbicides can render a normally unpalatable plant attractive to herbivores, but the inherent toxicity remains unchanged and poses a direct threat to the health of the animals (Mellanby 1967; Way 1969).

As a result of the widespread military use of 2,4,5-T as a defoliating agent, studies have been instituted to discover the cumulative, chronic effects of its presence in the environment. Preliminary evidence shows that impurities (chlorinated dibenzo-p-dioxins) that occur in commercial 2,4,5-T induce embryonic deformities and mortality in chicks (Risebrough et al. 1970).

With increasing use of herbicides in agriculture, in wildlife management, and in domestic gardens, we cannot disregard the possibility that there may be as yet undetected changes in wildlife populations (Persson 1968; New Scientist 1970, 49:593). While chronic effects may be unimportant in domestic or laboratory animals, they can appreciably affect the viability of a truly wild individual or species. Anything that reduces its competitive ability even for a limited time may make it vulnerable to other constraining factors in its environment.

To reiterate the point made earlier, when assessing damage to plant and animal life in any habitat, consideration must be given to the whole complex of environmental factors. An organism living in an urban environment is especially likely to be exposed to pollutants. The effects of pollution will differ according to the age of the organism, its physiological condition at the time of exposure, the frequency of such exposure, and the organism's physical environment, particularly the nature of the local climate. Pollution may cause ecological disruptions such as reduction in the number of species in an area. We are extremely ignorant of the full impact of the urban environment on living organisms beyond the point of knowing that an urban setting is inhospitable to a wide spectrum of species.

CHAPTER TWO

Characteristics of Urban Flora and Fauna

THE RESPONSE OF ORGANISMS TO THE URBAN ENVIRONMENT

Many of the animals that have been designated "pests" commonly possess the ability to adapt to changing conditions. Species possessing a wide range of tolerance to various environmental factors are most likely to be capable of co-existing with man. These plants and animals are often able to reproduce rapidly and to take advantage of transitory conditions or to evolve varieties suited to the urban situation. Animals with flexible behavioral patterns which enable them to exploit new food sources, new nesting or roosting sites, and to avoid novel forms of predation, may also be expected to colonize city habitats. Such intrinsic behavioral plasticity is possessed by many mammals and birds, allowing adaptation to occur within the lifetime of the individual without actual genetic change. Titmice (*Parus* spp.) are good examples of species with an ability to learn from observation, and it is possible for innovative behavior to spread rapidly throughout local populations.

Discovery of novel sources of food is well exemplified in the case of the British blue tit (*Parus caeruleus*) which in the 1940s developed the habit of stealing cream from milk bottles (Fisher and Hinde 1949). Similar reports have come from Norway where the great spotted woodpecker (*Dendrocopus major*), jackdaw (*Corvus monedula*), magpie (*Pica pica*) and blue tit open milk bottles (Abro 1964), a type of behavior that may result from an inborn tendency to hammer at hollow objects. In the United States the common crow (*Corvus brachyrhynchos*), blue jay (*Cyanocitta cristata*), and mockingbird (*Minus polyglottos*) are all well known for being able to take advantage of novel and relatively inaccessible food sources.

Some of the smaller diurnal mammals have become tolerant of people to the point where they may take food from the hand (Lond. Nat., 1968, 47:33). The gray squirrel in city parks has been tamed in this way yet still retains the essential degree of wariness necessary to protect it from danger. Indeed, it has been so successful in this habitat that some municipal authorities (New York for example) classify these animals as pests.

Very few animals except the common or Norway rat (*Rattus norvegicus*), black rat (*R. rattus*), house sparrow, and feral pigeon are found only in close association with human settlements. The ability of the sparrow to follow wherever man establishes new roots is demonstrated by the rapid spread of this bird following its introduction to America in 1850. Summers-Smith (1963) has discussed in detail the characteristics and achievements of the house sparrow in colonizing urban areas. The pigeon's commensal relationship with man has evolved over several centuries and Gompertz (1957) has suggested that the feral pigeons have merely exploited changed human behavior rather than having undergone change themselves. Their rock dove ancestors live on cliffs using caves for nesting, and fly to adjacent farmed areas to feed. The feral birds live on man-made cliffs and utilize scraps and refuse which accumulate in the city habitat (Goodwin 1954; Murton and Westwood 1966).

When organisms achieve pest status it is not always the result of adaptation. Newman (1966) noted that ants have often survived and developed noticeable populations because their predators have proved much more sensitive than the ants to the built environment. This probably applies to other insect species.

Animals least able to co-exist with *Homo sapiens* are his direct competitors, the carnivores. These mammals are the first to be driven away or exterminated when an area is settled. The larger the animal, the more likely that conflict will be intense; the limited distribution of animals such as bears (*Ursus* spp.), wolves (*Canis lupus*), cougars (*Felis concolor*), and lynxes (*Lynx canadensis*) exemplifies the result. Few carnivores can tolerate high levels of disturbance; they are driven from their haunts as peripheral sprawl occurs. They are most likely to survive in those situations where housing encroaches upon areas of sharp relief thus creating an abrupt juxtaposition of the built and natural environments (see chapters 3 and 4). What could be described as the medium-sized carnivores are very successful in the peripheral suburbs; typical species are the English fox and the American coyote (*Canis latrans*) (Teagle 1967; Gill 1970).

Another reason for the survival of some animals in the city is their nocturnal behavior. An omnivorous habit also bestows an advantage, allowing animals to exploit the wide range of food resources readily available in the city.

EXPLOITATION OF MAN-MADE STRUCTURES

Animals

Buildings are used by animals primarily for raising their young and roosting. A few species such as some insects and rodents may live their entire lives within buildings; however, most utilize them intermittently, since they require other habitat areas for feeding.

Starlings have adopted city centers as primary roosting places, and as a result in major capitals of Europe and North America they have achieved pest status. In London, the first indications of this change of roosting site were recorded in 1894, when starlings were seen gathering in the trees of St. James's Park (Johnson 1930). As the number of individuals increased, a movement from the trees to the ledges and sills of adjacent buildings was witnessed. To-day considerable expense is incurred by municipalities trying to dissuade the starlings from using buildings and in clearing up the results if they do roost there. The city roost has several advantages over a rural one: night-time temperatures are higher, there is an enormous footage of suitable ledges, and these high perches offer safety from predation.

Bats which originally roosted in hollow trees or caves have always found buildings quite acceptable substitutes, particularly favoring church spires and bell towers. Dilapi-dated houses with attics are often found to harbor large numbers of bats, which occupy them all year round (Hancock 1963; Beames 1968). Some gray squirrels in London have taken to living in the roof spaces of houses during the winter and similar records exist for Toronto (Lond. Nat., 1965, 44:54; Cox 1970).

Pied wagtails (*Motacilla alba*) have recently invaded the Hammersmith area of London where they roost on build-ings in flocks of some 120 individuals (Lond. Bird Rep., 1970, 34:58), and four hundred wagtails roost at the sewage purification works in London's Lea Valley (Glad-win 1963).

Roofs, parapets, and ledges of city buildings have served as nesting sites for many birds, including kestrels (*Falco tinnunculus*) in London (Lond. Bird Rep., 1968, 32:84), peregrine falcons in Montreal, Chicago, New York, Boston,

Philadelphia and Harrisburg (Groskin 1952; Hall 1970), sparrowhawks (*Falco sparverius*) in Washington, D.C., woodpigeons in London (Peal 1965), lesser black-backed gulls (*Larus fuscus*) in Gloucester, England (Owen 1967), and feral pigeons in most cities. Clearly many cliff-dwelling species find the built environment a very acceptable substitute for natural habitat; indeed, some species appear to prefer it.

In the wild, holes are utilized by many avian species for nesting purposes; old timber is a prime source of such nesting sites, as are sandbanks, river banks, and caves. In the city none of these native sites occurs in any quantity; thus birds are forced to adapt to different structures if they are to survive. Urban birds repeatedly nest in holes in the metal framework of power stations, gas holders, railway stations, warehouses, and other industrial sites. Timber yards provide excellent habitat, as do ruined or bombed buildings, quarries and gravel pits. In London the black redstart (*Phoenicurus ochruros*) has exploited most of these structures and reared offspring with increasing success in the last thirty years (Meadows 1965, 1970). In England, blackbirds, feral pigeons, and house sparrows, using metal girders as though they were tree branches, frequently nest in railway stations, bus terminals, and other public places. Magpies and kestrels have been seen nesting in the framework of a construction crane, and despite continual operation they fledged young without undue losses (Fairhurst 1970). House sparrows regularly use openings in neon display signs, lamp standards, utility poles, bus stop signs, roof eaves, drain pipes, ventilator hoods, and bridges. Other birds that have availed themselves of these artificial sites include swifts, woodpeckers, sand martins (*Riparia riparia*), titmice, English robins, and starlings. Gooders (1968) describes the house martin as breeding predominantly on man-made structures, particularly under roof eaves and bridges.

In North America similar adaptations are common, as when phoebes (*Sayornis* spp.) nest under eaves, barn owls (*Tyto alba*) in towers and abandoned buildings, and barn swallows (*Hirundo rustica*) in garages. Fischer (1958) has described the successful nesting of the chimney swift (*Chaetura pelagica*), which, in days before man built houses with chimneys, frequented caves, rock crevices, and hollow trees.

Persistent use of man-made structures is much less common among mammals, although there are reports of foxes using old cars as artificial dens (or "earths"), and in suburban England foxes use old sheds, shacks, drains, scrap yards and dumps for the rearing of cubs (Teagle 1967). Rats and mice are the only mammals that have adopted buildings as their principal habitat.

Plants

Many species can adapt to life on walls, which provide an important substrate for native plants in some urban areas. Thus, Kent (1961) notes that in Cambridge some 186 species have been reported growing on walls; in Durham there were 168; and his own work in Middlesex showed 204 species. In other areas of London, 70 walls supported some 83 species. An interesting aspect of this colonization is the high proportion of species long since banished from the surrounding area. Thus old walls may often host many locally rare species. Limestone memorials in churchyards and asbestos cement roofs both provide suitable alkaline substrates for lichens in the London area (Laundon 1970).

EXPLOITATION OF URBAN FOOD SOURCES
BY BIRDS

Adaptation to new resources is commonly displayed by birds in urban areas (Harrison 1960; Bergman 1961; Brown 1964; Williams 1964; Beven 1965; Yalden and Jones 1970). Studies in England have shown that tawny owls (*Strix aluco*) in rural areas have a diet comprised principally of small rodents (shrews, field mice, and voles), with birds contributing only 10 percent or less. Their urban counterparts by contrast, unable to find large quantities of small mammals, prey heavily upon local sparrows, starlings, and pigeons. The relative proportions for London owls is 93 percent birds and 7 percent rodents (mainly rats and house mice); such reversal of diet has also been confirmed in Paris and Manchester (Guichard 1957).

Other predatory birds living in inner urban districts are kestrels and peregrine falcons which, from the scanty evidence available, appear to be dependent upon the common passerines (Groskin 1952; Green and Forsyth 1970). Ljunggren (1968) observed that the eggs and young of urban wood pigeons in Sweden were less subjected to predation from carrion crows (*Corvus corone*) than those of their rural counterparts, a fact which he believed to be related to the numerous alternative sources of food available to the crows.

The readiness with which birds take household scraps and the more recently offered "wild" bird foods in garden feeders indicates that these are highly acceptable energy sources. Murton and Westwood (1966) analyzed crop contents of feral pigeons in Leeds and found that bread, cake, cereals, and exotic grass seeds were the principal components. Ljunggren (1968) and Cramp (1968) have confirmed the importance of bread as a dietary component for city wood pigeons in both Sweden and England. Such food may be inadequate, however, during the breeding season.

Titmice appear to utilize city food during the winter, and return to their natural woodland habitat to raise their young on the protein-rich and abundant caterpillar populations (Perrins 1968).

The growth in number of sewage outfalls, garbage dumps and other waste ground feeding sites has resulted in a concomitant increase in the populations of certain species throughout Europe and North America. Many gulls have become winter residents in urban areas and some remain to breed in central London (Meadows 1961; Sanderson 1968; Sage 1970). Gibbs (1963) studied rubbish dumps as a wildlife habitat and lists the principal species in the London area as gulls (*Larus* spp.), crows (*Corvus* spp.), the house sparrow, starling, feral pigeon, kestrel, and pied wagtail. Flocks of gulls in excess of 4,000 were seen at the larger dumps. Kestrels preyed on the large number of rodents present. In New York City and other northeastern American cities herring gulls have increased near sewage outlets and dumps (Bull 1964).

Strawinski (1963) pointed out the attractiveness of meat-packing plants to members of the crow family and noted the gatherings of seed-eating birds around warehouses, docks, and food-processing factories in Poland. Pigeons commonly frequent docks to feed upon spilled grain and manage to find adequate nesting sites inside warehouses. Sparrows exploit similar resources and are reported to be resident in bakeries, food processing plants, cafeterias, and canteens (Thearle 1968).

It seems clear that man's wasteful practices have encouraged some birds (and animals with omnivorous habits) to colonize urban areas. In their wake have come those predators whose behavior is sufficiently flexible to allow them to accommodate to noise, disturbance, and pollution. McNab (1963) and Schoener (1968) suggest that omnivores require a much smaller feeding area than do carnivores when compared on a body weight basis. This may also explain the incidence of colonial nesting among many omnivorous birds. In an urban area where the mosaic of habitat tends to be on a small scale it is not surprising to discover that a large number of omnivorous species can be found. In general there seems to be a high frequency of ground-feeding species among urban birds. Ground-nesting species, however, are rarely successful in urban areas.

MORTALITY IN URBAN ANIMALS

The relative importance of the different causes of death in urban animals is not known. Table 2 provides a list of various causes of mortality and the principle species affected in London.

One of the principal causes of animal mortality in cities is collision with various of man's artifacts. Ever since man first developed ways to transport himself at speeds faster than his own two feet would carry him, he has presented an increased danger to animals, whether or not he was actually using the conveyance to hunt them. In towns, cars more commonly injure humans, but in the suburbs and on roads through large parks many small mammals and songbirds meet with fatal collisions. High-impact areas have been discussed by Komarek and Wright (1929); Scott (1938); Pickles (1942); Finnis (1960); Hodson (1960, 1962); McCann (1960); and Dunthorn and Errington (1964). Breaks in the boundaries along roads often present a hazard for birds which tend to swoop low over the road surface where they become targets for fast moving vehicles. Birds and animals are drawn to railway rights-of-way by grain spilled from cars, exotic grasses which become established, and by the open nature of these rights-of-way. In isolated urban areas close to wild country, trains also claim substantial numbers of larger ungulates.

Table 2. Causes of Death in the Urban Environment (London, U.K.)

Mode	Animals Affected
Impact	
Vehicles—cars, trucks, buses trains, and aircraft	Foxes
	Hedgehogs, squirrels, weasels
	Rabbits, deer, songbirds
	Snakes, insects
Structures—glass windows, TV masts, tower buildings, power lines	Songbirds, particularly migrants
	Waterfowl, owls
Drowning	
Sewage canals and tanks	Hedgehogs
Ornamental pools	Owls, squirrels
Rainwater pipes and gutters	Sparrow chicks, sand martin chicks
Electrocution	
Railway lines	Badgers, rabbits, otters, foxes
Power transmission lines	Raptors, swans, geese
Neon display signs	Sparrows
Poisoning	
Pesticides	Foxes, badgers
Baits	Hedgehogs, birds
Heavy metals	Invertebrates, rodents, songbirds
Confinement	
Litter—cans and bottles	Small rodents, insects
Netting	Blackbirds
Predation	
Cats, dogs, man	Rodents, rabbits and hares, foxes, reptiles
Kestrel, tawny owl	Songbirds, amphibians
Crow, magpie, jay, jackdaw	Pigeons, doves
Fox, weasel, stoat	Waterfowl

Based on a survey of reports issued by the London Natural History Society, 1960-1970

In the London area there have been reports of high casualty rates among hedgehogs (*Erinaceus europaeus*) whose natural reaction to danger is to curl into a ball and remain still. Table 3 indicates a very minimum estimate of deaths in the London area due to this originally adaptive behavior. Naturalists in Lancashire, England, believe that hedgehogs are now showing signs of adapting to this hazard (Lancs. Nat. Trust 1970). A new strain has emerged that runs to the edge of the highway upon the approach of a vehicle, instead of curling up.

Among other animals that regularly succumb on roads are gray squirrels, weasels (*Mustela nivalis*), foxes, snakes, and innumerable insect species. Impact with static man-made structures is a problem for all avian species. A recent study of mute swans (*Cygnus olor*) in England, showed 40 percent of deaths were due to collisions with overhead wires (Brit. Birds 1970, 69:43).

Another source of mortality is poisoning, but the true effect of poisons and pesticides is unclear as there are few

Table 3. Hedgehog (*Erinaceus europaeus*) Mortality
on London Roadways

Year	Number Observed
1960	62
1961	50+
1962	66[a]
1966	199
1967	277[b]

[a] Of this number 26 were reported from a single stretch of road one mile long.
[b] Of this number 100 were seen on a 20-mile stretch of road in Essex and 11 came from a one-mile piece of road in a 12-week period.

Compiled from reports received by London Natural History Society 1960–1967 records of death directly attributable to this cause. Only occasional reports are found, such as that of a suburban gardener in London who found three dead hedgehogs in his flower bed after roses were sprayed with fungicide (Lond. Nat. 1969, 48:42). Teagle (1969) believes that badgers (*Meles meles*) in the London suburbs die from the toxins accumulated in their favored prey, the earthworm. Many small rodents die trapped in bottles and cans discarded in roadside ditches (Harper and Morris 1965).

Large numbers of cats and dogs to some extent occupy the vacuum created by the exclusion of native carnivores in cities. It is difficult to establish what degree of predation these domesticated animals exercise on urban wildlife, but the amount of predation is undoubtedly related to the number of feral individuals in any city. The well-cared-for household pet is unlikely to exert any marked predatory pressure on rodents and birds in its locality, but a neglected or truly feral cat has considerable influence. McKnight (1964) and Beck (1973) have reviewed the subject of feral animals and provided evidence concerning both the magnitude and diet of stray cat and dog populations in the United States.

Each year thousands of cats and kittens are abandoned, thus creating a substantial feral population. Some drift into rural areas or are purposely "lost" in the country, but many live within the urban environment and utilize garages, sheds, vacant lots, and sewers as habitat. The food of feral cats has been studied only in rural areas but is no doubt somewhat similar to that in urban locations. Rodents, particularly rats, are the principal dietary component, with garbage and birds contributing only a minor portion of the total intake; rabbits are important items at certain seasons but may well be absent from the urban dweller's diet. Cats seem to be a potentially significant influence on rodents throughout the year in all locations, and on songbirds during the breeding season in suburban districts.

Most reports on the food habits of feral dogs relate only to rural animals. Feral dogs however, have been recorded living in city slums, suburban parks, and in river valleys, particularly where these areas serve as refuse dumps. Known dietary items for feral dogs include muskrat, waterfowl, rabbit, squirrel, pheasant, opossum, and raccoon. An individual dog is less significant as a predator than a feral cat, however, since dogs more readily resort to scavenging.

Recently in Baltimore the first study of free-ranging dog ecology was made. Using standard techniques, Beck (1973) calculated the stray dog population of that city to be between 450 and 750 individuals per square mile. The total dog population in Baltimore was estimated to be between 65,000 and 162,000 individuals. Habitat commonly used by these dogs includes alleys, suburban shrubbery, junk yards, deserted buildings, and rooftops. Domestic garbage provided the principal food for stray dogs, but regular feeding by city residents was also significant. Home range of an individual dog was estimated as one-tenth of a square mile. Packs of two to six dogs were regularly seen. Beck concluded that these dogs were already becoming a problem, that their population was rapidly growing, and that some controls would soon have to be exercised.

Jackson (1951), working in residential areas of Baltimore, noted that an average cat caught between twenty-five and thirty rats each year and this had very little apparent effect on the population size of the prey. There is some evidence that a positive correlation exists between the rat population and that of dogs and cats; an area able to support feral cats can equally favor rats (Peters 1948). Backyard kennels that house domestic dogs may often attract rats, since the adequate supply of food and shelter adds suitable habitat. The population of dogs and cats in most cities is obviously large and is therefore an important component of the city ecosystem, and worthy of more study than it has yet received.

Other frequently observed predators in European cities are the carrion crow, tawny owl, fox, kestrel, and weasel. In North American cities, coyotes, striped skunks, and raccoons (*Procyon lotor*) can be added to the list. Man is less often a direct predator upon animals in the city than he is in rural areas, but children do represent a threat to animal populations in local areas. The use of air guns, slingshots, and the practice of egg collecting can all raise the level of predation so that it more closely resembles that experienced by rural populations (Yalden 1965).

When does mortality occur? For animals in natural habitats the heaviest losses take place toward the end of the winter period as food supplies become exhausted or increasingly inaccessible. Urban-suburban fauna are less affected by the deprivations of winter, and the heaviest adult mortality is found in the breeding season. Snow (1958) compared English blackbird populations living in woodland and suburban parkland and observed a lower adult mortality rate among the garden birds during months of winter. This altered seasonal mortality has also been confirmed by others (Havlin 1963; Erz 1964).

The adults of most urban-suburban songbird species die predominantly in the breeding season (April to June) while brooding or during their search for food to satisfy the nestlings, when it is believed they become less wary (Summers-Smith 1956; Snow 1958; Coulson 1961).

Grey squirrels living in Toronto city parks appear to show a decreased winter mortality and again the shift is into the months of April to June (Cox 1970). Information on other animals is yet to be collected but it is reasonable to expect that the advantageous food resource and warmer winter temperatures of the city will produce a similar shift in the period of high mortality.

PHYSIOLOGY AND BEHAVIOR OF
URBAN - SUBURBAN MAMMALS

This is an area in which few scientific studies exist. Casual observation is the major source of information; thus an interpretation can be only speculative.

Vesey-Fitzgerald (1965) suggests that suburban foxes in southern England have become noticeably diurnal, extending their activities into the late morning; they may regularly be seen loping across gardens on their way back to dens from their night-time hunting forays. These same foxes appear to be breeding up to three months earlier than rural ones. Presumably increased food combined with freedom from persecution have made such changes possible. To illustrate, Vesey-Fitzgerald believes that suburban vixens do not bother to move their litters to a different den shortly after birth since such behavior is adaptive only when the animal is regularly hunted. In gardens of private homes the vixen is less likely to be disturbed; there may also be a shortage of fox dens in suburban areas.

Cox (1970) studied the gray squirrel in Queen's Park, Toronto, and concluded that human presence had caused changes in the animal's periods of peak activity. Lunchtime strollers usually fed the squirrels and certain local residents came each day throughout the year specifically for the purpose of feeding them peanuts. He noted that in the wild habitat, gray squirrels were observed to be most active during the three-hour periods following sunrise and before sunset. Toronto squirrels displayed in additional noon-hour period of activity during which paths were the principal focus areas.

BEHAVIOR OF URBAN-SUBURBAN BIRDS

Both American and English robins become very tame and tolerant of human presence. Many gardeners are familiar with this bird, which closely attends any digging activity in order to capture exposed worms and larvae. A more specialized behavior was reported from Holland Park in London where a robin, whose territory included a log pile, would immediately fly to the pile when a person approached and wait for a log to be turned over so that he could snap up the exposed invertebrates (Brown 1963).

From Poland, Dyrcz (1969) has discussed the predominance in built-up areas of the blackbird over the song thrush (*Turdus philomelos*) and believes this to be the result of the former species' greater adaptability. It is able to feed upon a wider range of materials and also accepts a variety of alternative nesting sites when living in areas devoid of shrubs.

There seems to be no evidence that new behavioral patterns emerge in the urban environment; there is instead a persistence of old patterns which are directed toward new resources (e.g. the titmice and milk).

POPULATION CHARACTERISTICS OF
URBAN-SUBURBAN BIRDS

More has been written on birds in cities than on any other aspect of urban ecology. The mobility of these animals makes them better suited to the urban environment than most other forms. Work in North America, Europe, and Britain suggests that many songbirds are actually more successful in a suburban than in a rural habitat; in well-established residential districts even species diversity is moderately high (Strawinski 1963; Woolfenden and Rohwer 1969).

The density of urban birds appears to depend upon the type of vegetation cover available. Areas where stratified vegetation, buildings, arable and pasture land are interspersed with plentiful water sources can support a varied,

Table 4. Man-Modified Habitats and Songbird Densities

Habitat	Breeding Density (pairs/10 acres)	Location	Source
Urban-Suburban			
Parkland	185.0	Wisconsin	Young 1949
Parkland	141.5	France	Ribaut 1970
Campuses, estates and parks	av. 41.4	U.S.A.	Odum 1959
Sportsground	34.4	France	Ribaut 1970
Gravel pit	24.8-45.0	England	Glue 1970
Inner City parkland	24.0	London	Sanderson 1968
Town cemetery	72.0	Switzerland	Ribaut 1970
Old suburban district	26.0	London	Simms 1962
Old suburban district	av. 55.0	Florida	Woolfenden and Rohwer 1969
Suburb built in former park	49.4	Zeist, Holland	Bruyns 1961
New suburban district	28.2	Dortmund, Germany	Erz 1959
New suburban district	20.0	Florida	Woolfenden and Rohwer 1969
Rural			
Parkland	28.8	Poland	Havlin 1963
Parkland-Farmland	44.6	England	Ribaut 1970
Parkland-Farmland	27.5	France	Ribaut 1970
Woodland (oak)	53.5	Scotland	Williamson 1969
Woodland (birch)	44.5	Scotland	Williamson 1969
Woodland	30.3	Switzerland	Ribaut 1970

high-density population throughout the year. Table 4 shows the range of densities in different locations. It is clear that geographical location fundamentally influences the potential density, but more examples are required before definite statements can be made concerning relative densities of urban and rural locations. Using the English blackbird as one example, however, it does appear that this bird attains its greatest breeding density in peripheral, residential districts of cities (Table 5). In some instances densities may be as much as four times those of rural woodland populations (Havlin 1963).

In contrast to these figures, the evidence thus far available shows that suburban European titmice, blackbirds, song thrushes, robins and dunnocks (*Prunella modularis*) lay smaller egg clutches than members of the same species inhabiting woodlands. The smallest clutches recorded are those from the inner city. Clutch size, of course, generally reflects the quality of environmental conditions (Snow 1958; Havlin 1963; Lack 1965; Snow and Mayer-Gross 1967; Perrins 1968).

Songbird nesting success depends upon a complex of factors of which habitat type is most important. Birds that habitually nest early and build noticeable, bulky nests are extremely vulnerable in urban situations where little understory vegetation exists. The earlier onset of the growing season in the warmer urban environment may help to explain the observed earlier nesting of several songbird species. Where cover is available in the city, nesting success is usual; in England, Snow and Mayer-Gross (1967) report greater success of blackbird and song thrush breeding in urban gardens than in woodlands.

In considering urban-rural songbird nestling mortality, it has been recorded that great tits (*Parus major*) experience a higher nestling mortality in the city than do rural populations (42.5% vs. 33.0%). There is also a preferential sur-

Table 5. Population Characteristics of Blackbird (*Turdus merula*) in Different Locations

	England large suburban garden	Czecho-slovakia suburb	England dense oakwood	England farm-land	Czecho-slovakia woodland	England urban area
Breeding season	1–2 weeks earlier than in woodland	10 days earlier than in woodland				
Density of birds (Pairs/10 acres)	20.0	4x that in wood-land	2–3	1.0		
Clutch size	3.8	3.94		3.91	4.14	
Nesting success	50%	69%	12%	28%	62.4%	
Fledging success	83%		89%	81%		73%
Average number of young raised per nest	2.1	2.4			2.46	

Compiled from Snow 1958; Havlin 1963; Snow and Mayer-Gross 1967.

vival of male nestlings; thus urban populations exhibit a distorted sex ratio (Dhondt 1970), apparently a consequence of insufficient feeding which increases the competition between the heavier-bodied male nestlings and their female siblings. This adds substance to the comments made elsewhere concerning the probable shortage of high-protein food necessary to achieve satisfactory fledging success. Only those species capable of utilizing new or different protein sources can be expected to enjoy comparable success in the urban and rural environments.

Food may often be the limiting factor which reduces the fledging success of urban-suburban birds. Blackbirds, for whom the earthworm is a major dietary component, seem able to find sufficient food in gardens and lawns during the early part of the breeding season, which allows them to raise three or four offspring. The success rate declines as the ground dries and less open ground is exposed. Titmice, which tend to be arboreal feeders, can rarely raise the eight or ten young commonly found in their native habitat. In the suburbs extensive woodland capable of producing the prolific caterpillar crop necessary for such birds is very rarely present (Perrins 1968).

The other important variable is the rate of predation, including nest destruction. Again, it appears that urban songbirds may suffer less damage to their nests and thus potentially higher nesting success is possible (Snow and Mayer-Gross 1967). The prime mortality during the breeding season is suffered by adults, probably as a result of the large amount of time spent searching for food on the ground where they can easily fall prey to cats and other predators.

If the blackbird (which has been the most studied) is typical, many suburban areas seem to provide a surplus of individuals which may either disperse and colonize new areas of built-up districts or return to more rural areas where lower densities pertain.

There are few species other than songbirds found in the city, and there are few that originally derive from habitat other than woodland. Birds typical of open grassland are poorly represented in the built environment, as are marshland species. Cliff-dwelling species such as the feral pigeon are prolific in the city, because they breed almost year round in response to elevated temperatures and a plentiful food supply. The unique study of the peregrine falcon in Montreal indicated that this species may also enjoy heightened success when raising young in the urban environment (Hall 1970). Hickey (1970) reported an average success rate of 1.1 fledglings per nest site in natural mountainous habitat which can be compared to Hall's figures for urban falcons, of 2.2 per nest over the period 1940 to 1952.

CHAPTER THREE

London: A City with Integrated Suburban Wildlife Habitat

Since 1858 London has had an active natural history society (LNHS) to record its many species of indigenous and exotic flora and fauna. Until recently the emphasis has been on identification and accumulation of species numbers, but ecological studies have emerged in the past decade. Table 6 has been compiled from LNHS annual reports for the years 1960–70 and provides evidence of the viability of metropolitan London as a "wildlife" area.

It is highly probable that many cities are equally well endowed with plant and animal species; but, due to a scarcity of urban naturalists, little documentation exists and thus the true diversity of the city is unknown to most of its residents.

NATURAL HISTORY OF LONDON

It will be apparent from Table 6 that a rich variety of species persists in the metropolis along with its eight million human residents. That some animals have been less able to remain than others is shown by the relatively few carnivorous species recorded. Only the smaller carnivores

Table 6. Flora and Fauna of the London Area, 1960–1970

Type	No. of Species		
	Seen in London	Seen in UK	Per cent of UK Species Present in London
Terrestrial Mammals	32	52	62
Insectivora	5	6	83
Chiroptera	10	15	66
Carnivora	4	11	36
Artiodactyla	3	9	33
Lagomorpha	2	3	66
Rodentia	8	17	47
Reptiles	6	10	60
Amphibians	8	12	66
Fishes	33	45[a]	73
Birds	203	301	67
Invertebrates			
Macro-Lepidotera	728		78
Hemiptera-Homoptera	317		82
Diptera	2,300	5,200	44
Coleoptera	248	3,700	7
Plants	1,835	ca.3,000	61

The London area is described as that contained in a circle of radius twenty miles from St. Paul's Cathedral, approximately 1200 square miles.

[a]Fresh and brackish water fishes only.

Compiled from: London Natural History Society reports 1960–70; Fitter 1963; Southern 1964.

are tolerated and in Britain as a whole very few of the indigenous species of carnivores still remain (Southern 1964). By contrast, herbivores, even the larger ones such as deer, are still to be found in cities although they are not truly wild. Three semi-wild herds exist in London: one at the Royal Park in Richmond, another at Bushy Park, and the third in Epping Forest (Hurcomb 1969).

Smaller animals, particularly invertebrates, thrive in large numbers wherever suitable habitat has been retained or created. A good example of a relatively undisturbed area which supports a highly varied plant and insect population is the garden of Buckingham Palace (McClintock 1964; de Worms 1965). Some 14 per cent of all Lepidopteran species found in the British Isles have been reported from this central oasis.

Resident birds are the most conspicuous of all city animals. Fitter (1945) listed 14 species which he termed the "garden association" as being those most successfully adapted to living with man. As a result of its location, London also serves as an important staging area for migrants passing through in spring and autumn (Lond. Bird Rep. 1963, 27:5). Certain of the central London parks regularly report unusual species in times of bad weather conditions on the Continent (Min. Pub. Bldg. and Works 1970). There are few raptors in the resident avian population but the kestrel is still found in the inner districts and tawny owls thrive wherever they can find adequate nesting and roosting sites (Brown 1963; Montier 1968).

Many plants are introduced ornamentals but a wide selection of the so-called weeds are present in parks, gardens, cemeteries, and on waste ground. With the large bird population and the abundance of seed sources within and surrounding the built area there is unlikely to be a shortage of propagules to exploit any freshly turned soil (Peterken 1953; Bangerter 1961).

In order to attempt an understanding of the underlying reasons for the maintenance of such a rich natural element in the metropolitan landscape, it is necessary to select the probable controlling factors and examine them individually. An excellent review of the natural history of London was published in 1945 by Fitter; this, together with the aforementioned reports of the LNHS, provides principal information sources.

CLIMATE

According to Köppen's classification of climate the whole of the British Isles is included in the warm, temperate zone which does not experience a marked rainy season. The climate of the London area has been treated in detail by Chandler (1965). Essential features of the regional climate include a relatively wide range of monthly mean temperatures in the London basin (36°F range, from 35°F to 71°F) mainly as a result of the higher mean summer temperatures when compared to the other parts of Britain. Cold winters are a rarity and snow cover is expected only five to ten days in an average year. Rainfall (annual average 21–29 inches) is distributed throughout the year and a relatively high humidity persists most of the time. The prevailing winds are from the southwest and the orientation of the Thames Valley in the London area results in inefficient dispersion of air pollutants in the low-lying industrial areas.

In common with all cities, London suffers air pollution problems, the important pollutants being sulphur oxides and smoke. Following the passing of the Clean Air Act of 1954, a number of smokeless fuel districts have been created and tangible improvements in air quality documented. Seasonally high accumulations of pollutants nevertheless continue. Commins and Walker (1967) calculated that over the ten years 1954–1964 smoke concentrations were three times greater in the winter than in summer. Similarly, the sulphur dioxide concentrations were twice as great in winter. According to Chandler (1970) smoke emissions have decreased by 50 per cent, and sulphur emissions, while increasing by some 30 per cent, have not produced an increase in the ambient concentrations of sulphur dioxide at ground level in the city. Since 1956 the frequency of severe fogs has dropped and an increase in sunshine hours recorded between September and March (1958–67) is evident in central London (Jenkins 1969; Chandler 1970). Thus London has a mild climate which favors a wide variety of both resident and migrant species, and recent improvements in air quality may be expected to further increase its attractiveness as an environment for wildlife.

SPATIAL DEVELOPMENT

London was first settled by the Romans around A.D. 50 and in the past nineteen hundred years has grown from an original area of 330 acres (0.52 square miles) to its present 750 square miles. Since 1945 government policy has dictated that growth be restricted and much of the expansion in the region has occurred in the surrounding countryside where new towns have been located. The population density of London displays a pattern of concentric rings of decreasing density around the central district. (In the last decade the residential population of the central area declined markedly but there still remains a large daytime office population.) The urban area extends to a fifteen mile radius, beyond which a "green belt" preserves significant amounts of open countryside. This basic pattern is disrupted by radial developments along major transport arteries. With respect to land use the metropolis can be divided into three principal zones: a central area dominated by high-rise office blocks and shops, a belt of three to seven miles containing factories and houses built in the eighteenth and nineteenth centuries, and a peripheral belt of suburban twentieth century housing, largely of single family, low-density type. This outer land use sporadically penetrates the green belt, creating a juxtaposition of the natural and built environments (Clayton 1964).

VEGETATION IN THE LONDON AREA

The influence of man is so marked throughout the lowland regions of Britain that it is difficult to distinguish indigenous vegetation. Originally, the clay lands were thickly covered by oak (*Quercus* spp.) while the drier chalk uplands supported beech woodlands (*Fagus sylvaticus*). Areas with sandy deposits probably harbored a heathland flora but may well have had a scrub oak and birch cover in the undisturbed condition. None of this original cover exists today following the early invasion of the chalk lands for agriculture and the later clearance of the oakwoods (Darby 1951).

Remnants of mixed woodland dating from the eleventh century may still be found, but they tend to be much less dense than the original cover, and are comprised of mature trees scattered among young trees with a rich understory of shrubs and herbaceous species (Peterken 1953).

Aquatic vegetation can still be found in rivers and in ornamental lakes or pools such as those which adorn many city parks. Little remains, however, of the extensive marsh and bog vegetation which formed an important component in the pre-Roman landscape.

Turrill (1948) has suggested that few species have been extirpated from the British flora entirely as a consequence of human activities but many aliens have been introduced. Some of these exotic species such as elm, larch, and sycamore (*Acer pseudoplatanus*) were intentionally imported for their ornamental qualities, but many arrived in consignments of crops and other raw materials. Within the

British Isles a great deal of interchange of genetic material has resulted from propagules being carried to urban areas in local building materials.

NATURAL CONSTRAINTS UPON DEVELOPMENT

Within the Greater London area local relief ranges from 10 to 882 feet above sea level. The primary feature is the Thames basin with its famous river and associated gravel terraces and alluvial flood plain. The predominant direction of drainage is from west to east; in the west the river flows through river alluvium, clay, and gravels, whereas to the east of the city between Dartford and Gravesend it penetrates the chalk outcrop. It is tidal as far as Teddington, with a range of 20 feet near London Bridge.

The river valley was the site of the earliest construction, and settlement was limited to gravel hills, the outline of which may still be detected within central London. The expansion of building beyond the valley was delayed until the nineteenth century because the extensive London Clay to the north presented an intractable environment to the then contemporary technology. Tertiary deposits to both the north and south of the city provided land at higher elevation which became the preferred residential locations of the rich city merchants in the seventeenth and eighteenth centuries (Clayton 1964).

Thus London until the start of the nineteenth century was limited to those parts of the region which provided firm foundations, workable soils, and good drainage, and which allowed wells to be drilled for drinking water. Marsh land and clay soils remained much as they had been for thousands of years except for some low intensity grazing for dairying purposes. The extent of the settled area at that time was small, with a population of 900,000. After

1801 changes in technology resulted in a quadrupling of this population within eighty years; by 1880 the population was 3,680,000.

The terrestrial wildlife habitat of the region was not greatly diminished by the growth prior to the nineteenth century, and only central London, where air pollution and sanitation problems abounded, was inimical to life. The floral and faunal components of attractive uplands such as Hampstead Heath or Sydenham Heights were modified by the creation of country estates. The agricultural methods tended to produce a very diverse, small-scale landscape of fields and hedgerows which increased the species diversity in the associated animal community compared to that which existed in the dense oak woods. Industry was restricted to some clay workings and brick kilns which no doubt produced local air pollution but added numerous small ponds to the landscape.

The same could not be said for the aquatic habitat which even in the very earliest phases of settlement was subjected to considerable change. Butler (1962) has traced the demise of London's smaller streams and rivers commencing with the covering in of the Wall Brook in 1461 and the subsequent taming of the Langbourne, the Holebourne, the Tyburn, the Fleet, and many others.

The growing need for potable water saw the beginning of the gradual lowering of the water table (which still continues) and this has added to the loss of wetland areas. This era of limited, slow modification of the countryside came to an end around the middle of the nineteenth century when new skills released the human population from most of the controls exercised upon it by the landscape. Thus it was that the wetlands were drained and filled, the steepest hills were cut and leveled to allow building, and drainage systems using earthenware pipes brought about the surrender of the clay lands to the invading suburbs (Clayton 1964; Dancer and Hardy 1969).

With man's competitive powers so enhanced, the principal means by which wildlife habitat could be retained in the metropolis was by government regulation.

LONDON'S DEVELOPMENT:
THE LAST ONE HUNDRED FIFTY YEARS

The growth in the metropolitan population continued with only a temporary halt occasioned by the First World War. From 1921 to 1939 the population of London grew at twice the national rate. By 1961 the city region contained more than 13 million people; it is projected that the figure in 1981 will be close to 16 million. As a result of the adoption of the Abercrombie proposal for a Green Belt, a substantial proportion of the postwar growth has been in satellite settlements. The intervening land has been left in agricultural or recreational use, which has generally favored the maintenance of wildlife populations (Figure 1).

Pollution

An important outcome of the explosive growth in London's population between 1800 and 1940 was the rise in all forms of pollution. The levels were such that Thameside ports were unable to continue commercial fishing; the last salmon was taken in 1833 (Fitter 1945; Wheeler 1969a). The vegetation was assailed by dusts and sulphurous fumes from the many industrial and domestic furnaces burning soft coal. Although relatively few records were kept, many species must have been extirpated from the heavily populated areas (Fitter 1945; Bangerter 1961).

While planning legislation enacted in the forties managed to halt the continued peripheral expansion of the built area, it was not until the 1950s that the first comprehensive attempt was made to control pollution. In 1954

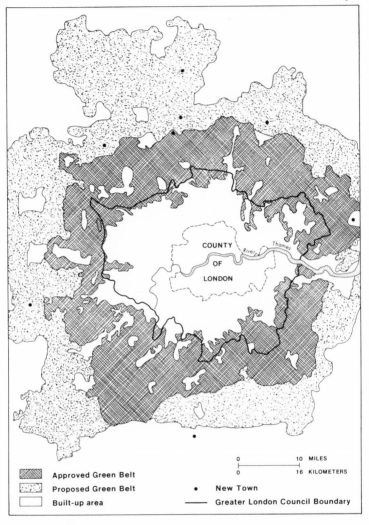

Legend:
- Approved Green Belt
- Proposed Green Belt
- Built-up area
- New Town
- Greater London Council Boundary

0 — 10 MILES
0 — 16 KILOMETERS

Figure 1. London and the Green Belt.

the Clean Air Act created "smokeless zones" which have decreased smoke emissions. The return of both the swift and the house martin to the inner London area is believed to be connected to this improvement (Cramp and Gooders 1967; Gooders 1968).

The Water Resources Act of 1963 has been instrumental in bringing about significant improvements in the quality of Thames water (Marlborough 1963; Wheeler 1969a, 1970). For illustration, in 1951-1954 a weekly sampling of the water between Putney and Southend showed that contamination was greatest in the summer months and resulted in noticeable amounts of hydrogen sulfide gas emanating from the river. Dissolved oxygen almost disappeared from the river water during such periods from five miles above London Bridge to a point some 27 miles below. However, in 1970 Wheeler was able to report that a total of 42 species of fish had been recently recorded in the Thames below London—an indication of the success of the pollution controls. An additional benefit attributed to reduced water pollution has been the presence during recent winters of large flocks of waders and waterfowl on the lower reaches of the Thames (Grant 1970).

The obvious forms of pollution have been diminished in the metropolis over the last decade while the less perceptible forms have grown in number and intensity. Motor vehicles have contributed greatly to this increase; so too has the accumulation of solid wastes and the various techniques employed in their disposal. The widespread use of synthetic, persistent compounds has resulted in these materials being dispersed throughout the global ecosystem and they can be assumed to be important factors in the viability of wildlife populations in London (Tarrant and Tatton 1968).

Habitat Modification

The landscape created by agricultural activity in the past centuries has in turn benefited and deterred numerous plant and animal species. Until the 1950s farming had produced a mosaic of small fields separated by hedgerows and lanes with an accompanying profusion of "edge" spe-

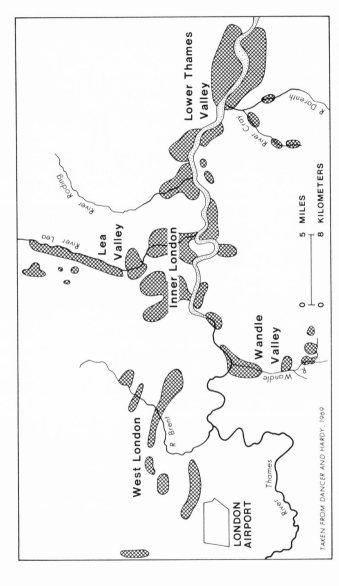

Figure 2. Concentration of Industrial Areas along Rivers. (From W. S. Dancer and A. V. Hardy. *Greater London.* Cambridge University Press.)

cies. The recent trend to large field units has resulted in a drastic reduction in the hedgerow habitat, to the detriment of some local wildlife populations (Hooper 1970). In the London region, however, the impact of this change has not been very noticeable, except on the eastern edge.

Land in the London area which has once been farmed and later abandoned shows a gradual return to a tree-dominated cover. This transition may be observed around the urban area wherever land values reach levels that make development economically attractive. Other areas in addition to the urban-rural fringe also show this phenomenon. Many tracts of common land which now lie within the London region (e.g., Bookham and Ham Commons) are no longer grazed by stock and are thus reverting to scrub and potentially to woodland. Many of these open spaces now serve as prime recreational sites and despite the release from grazing pressure their vegetation shows arrested development due to trampling and repeated fires (Lond. Nat. 1962, 41:23; and 1965, 44:126).

The impact of increased human numbers on a slowly decreasing amount of open space has meant a reduction in the usefulness of this acreage as wildlife habitat. Carrying capacity is affected when large areas are turned over to lawn and sports ground; dogs and children trample peripheral vegetation, and the use of these spaces extends throughout the year.

The advent of the automobile in the early twentieth century has had its effect through loss of habitat, increased pollutants, and as an instrument of death for small birds, insects, and mammals. Despite the acknowledged importance and uniqueness of the Royal Parks that grace the central area, not even these are inviolate when the needs of the car are to be satisfied. Hyde Park, in addition to many other parks and commons, has lost acreage in the race for extra parking space and highway extension to alleviate traffic congestion.

One consequence of increasing construction has been the extensive excavation of building materials such as limestone, sand, gravel, and clay. Initially such activity means destruction of plant and animal communities, but following completion of the extraction a new habitat is left which can be very valuable. Unfortunately many of the pits and quarries so created have been used to dump rubbish and other unwanted materials and the land later placed into use for recreation or grazing. Many, however, remain as unused land and can be important oases of wildlife. For example, chalk pits in the south and east sections of the London region are very rich botanical sites (Peterken 1953; Lond. Nat. 1967, 46:24-25).

For many years one of the most popular methods of garbage disposal was the filling of marsh along the Thames valley, with the consequence that the distinctive marsh ecosystem is now very rare in the London area and relegated to landscape remnants or to man-made structures which replicate to some extent the original conditions. Additional aquatic sites have been provided through the years by the excavation of canals, reservoirs, and gravel and clay pits. Gravel pits are often in flood plains and thus become man-made lakes which can simulate natural conditions attractive to waterfowl and small mammals (Glue 1970; Harrison 1970).

One other loss, occasioned by the growth in the number of single-family dwellings, is the demand for turf to give the innumerable small gardens their most characteristic feature. Some of the commons (Hampstead Heath, Mitcham and Wimbledon Common) have suffered very badly from these deprivations, losing turf, sand, gravel, and peat in the face of development.

The original housing developments in the countryside around the City of London took the form of mansions set in either parkland or woodland hunting preserves. Following the era of suburban sprawl in the early twentieth

century, these estates were surrounded and many suc-
cumbed to multiple housing development schemes. A small
number were acquired by far-sighted municipalities who
saw their amenity value and their potential use as "green
lungs" for the densely populated districts.

These estates had provided a high capacity woodland
habitat that acted as a reservoir from which migrants could
colonize less viable wildlife areas. Where these estates were
broken up their loss had marked effects upon local wildlife
populations.

World War II, 1939-1945

During the Second World War a macabre experiment in
habitat manipulation was enacted. The influence of the
London Blitz upon the human population is well known,
but it had some equally pronounced effects upon the
plants and animals of the area (Fitter 1945).

In the days between September 1940 and May 1941
approximately one-third of the old City of London was
devastated and in subsequent years other areas of the
metropolis suffered extensive damage. The outcome of
these sudden changes in the environment was the provision
of a diversity of small spaces comprised of a mixture of
rubble piles, ruined structures, and freshly turned soil. The
resilience of natural plant populations was amply demonstrated by the positive riot of annual species which soon
colonized the bombed sites. A survey conducted in 1942
revealed a total of 27 plant species on such sites, and this
count rose to 126 by the end of the war (Salisbury 1945).
The species involved were typical of pioneer communities,
distinguished by their ability to germinate and grow rapidly in transient and disturbed conditions. Where the sites
received no further damage normal succession occurred
with the result that a shrub community developed (Parmenter 1968). The earth which had been built over for

thousands of years proved to be productive and capable of furnishing wildlife habitat within a few months of being once again exposed.

An additional feature of wartime London was the omni-present, open-topped water storage tank, and these to-gether with the scattered pieces of meadow and scrub-type land allowed a gradual colonization by animals. Yet another aspect of importance in explaining the biological invasion which occurred at this time was the number of domestic gardens, many of which were neglected and be-came the equivalent of reserves or sanctuaries.

Ruined buildings furnished a variety of nesting and roosting sites which were taken advantage of by birds, insects, and rodents. It is likely that other mammals may also have been present, but it should be recalled that during this period few naturalists had free time and much evidence that might have been available was never re-corded.

A notable species that exploited the bombed sites was the black redstart. This bird had not been recorded as a breeding species in central London before 1940. The first broods were reared in the precincts of Westminster Abbey and by the year 1949 the tally in the central area had reached 14 pairs all successfully feeding on the insects that associated with the herbaceous ground cover. No doubt they also benefited from the presence of the water tanks, which must have been a rich source of flying insects in the warmer months of the year. Other birds enjoyed this newly provided resource area and it was reported that migrant warblers were to be seen searching the foliage on the older bombed sites where elder (*Sambucus nigra*) and willow flourished.

Some acreage in the larger parks was plowed and turned over to crop production during the war. There was also some felling of commercial quality timber in the London area, but the trees most valuable to wildlife were overmature and hence not attractive to the forester. All other evidence suggests that there was an appreciable increase in natural vegetation and reduced intrusion and disturbance by people. The devastation witnessed in London at this time, while greatly detrimental to man, was nevertheless instructive in revealing the capacity for other organisms to colonize the previously built-up environment.

To a lesser degree this same process was repeated in the early sixties when underpass and highway construction in Hyde Park produced fresh expanses of earth. Several observers reported a bewildering array of annual grasses and weedy species, noting particularly the occurrence of alien grasses. This was a new development not apparent during the wartime plant invasion. The source of such grasses is believed to be commercial bird seed which in recent years has been continually scattered by people feeding wild birds. Undoubtedly many seeds are also discarded into garbage bins and these ultimately reach open dumps where they may be transported by wind or scavenging birds (Lond. Nat. 1962, 41:13; 1963, 42:11; 1967, 42:26).

GREEN SPACE AND WILDLIFE HABITAT IN LONDON

Within the metropolis a great deal of unbuilt land has been retained and this serves a variety of functions including provision of wildlife habitat. Table 7 lists the principal types of land use and the form of natural habitat they provide.

In view of the statement made earlier concerning the lifting of restraints upon development during the nineteenth century, it is important to consider by what mechanism the present-day green space has been retained. Most of this land seems to have been saved by accident rather than as a result of far-sighted planning by the city fathers.

Table 7. City Habitats

Natural Habitat Type	Urban Land Use	
Wetland	Gravel pits Sewage farms	
Aquatic	Sewage farms Water tanks Reservoirs Gravel, clay pits Ponds and lakes in parks/gardens Watercress farms	
Open–grassland	Airports Playing fields School grounds Roadside verges Public parks	Common land Railway banks Canal banks Race courses
Open–arable	Gardens Agricultural and horticultural land	
Parkland	Institutional grounds–hospitals, prisons Public parks Cemeteries Golf courses Common land	
Woodland edge	Hedgerow Schools Institutional grounds Garden shrubberies Churchyards Golf courses	
Cliff and ravine	Chalk pits Quarries Walls Bridges Viaducts and aqueducts Railway cuttings	

Public Open Space

Parks in London are the property of the Crown, the Corporation of London, or the Greater London Council; the latter two bodies also own common land and other open spaces in the metropolitan region. The total land involved is of the order of 28,000–29,000 acres. (When added to the land area contained within the Green Belt a total of 40,000 acres is set aside as permanent green space for public use.) Not all of the land is open or green nor does it all provide wildlife habitat, since this total includes land upon which houses, hospitals, offices, and libraries stand, as well as areas such as sports grounds, golf courses, and stadia. Such land is nevertheless useful as a feeding ground for birds by day and may serve a similar function for nocturnal mammals. Many of these areas are linked to one another or to a corridor of some kind, and thus much of the acreage is able to support some wild species. The critical areas are those with good ground cover and mixed shrub and tree communities which can act as reservoirs to supply the less optimal locations with replacement organisms.

The principal central London parks have their origins in the medieval period when they were set aside as Royal hunting preserves or sites for country palaces. Over the years the people gradually acquired the right of access to these "noble playgrounds." In the mid-nineteenth century a "parks movement" blossomed and resulted in local governments acquiring large private estates and derelict land within their administrative areas for conversion to public parks or institutional use (Church 1956; Chadwick 1965; Trent 1965; Jellicoe 1970).

The parks which were designed for the people of London in the 1800s were very formal; no attention was paid to their potential use by wild species. It has only been in recent times that a maturity and naturalness has developed in these nineteenth century parks, making them very attractive to a wide variety of birds and a number of mammals (Chadwick 1966).

One other very characteristic form of green space which enhances the image of London is the delightful squares which date from the seventeenth, eighteenth, and nineteenth centuries. Some still remain enclosed, with keys available only to residents of the square while others such as Soho and Berkeley provide resting places for walkers, residents, and office workers alike. In terms of their wildlife habitat they hold an esthetic rather than a practical value since only a few species can exploit these small areas. Wood pigeons, crows, and even magpies have been reported nesting in the plane trees (*Platanus acerifolia*) of London squares.

The other important source of metropolitan open space is the peculiarly British feature known as a "common." Such land was the common pasturage of the peripheral villages which existed prior to the nineteenth century explosion of London's urban populace. Most of the commonland within the London area is now in the hands of local government and thus becomes legally open for public use. Unfortunately now such land can be swallowed by highway construction, as in many cases the owner of the land and the highway builder are the same.

These major pieces of countryside, which were subsequently to be locked in the urban fabric, must have been significant factors determining the location of high class residences. These houses with spacious grounds adjacent to the boundaries of the commons greatly enhanced the wildlife potential of these areas. The continued expansion of London and the rise in land prices has brought about the demise of many such elegant properties. The result is that the grounds have been fragmented and much of the land has disappeared under concrete and asphalt to provide car parking space. This has reduced the buffer zone around the commons and made them less viable as wildlife reservoirs.

One fortunate outcome of the historical derivation of central London's green space lies in its location. Standing as they do on Tertiary deposits, most of the major parks and commons are elevated above their surroundings. Such sites are favored with low air pollution levels due to stronger winds, increased precipitation, and greater sunshine (Chandler 1965).

The Open Space Network

The verdant aspects of Greater London are accounted for largely by the indulgent tastes of medieval monarchs, together with the actions of philanthropists who, in sympathy with the few far-sighted planners, started battling to retain greenery during the first decades of the nineteenth century. In a few cases the banding together of private citizens to purchase land by subscription has secured important natural areas.

The result of utilizing such a haphazard mechanism is the present abundance of isolated units scattered throughout the urban fabric. There is no obvious open space network as may be detected in say Cleveland, or in Stockholm, where definite plans of land acquisition were followed (Tunnard and Pushkarev 1963). Nevertheless, water courses and transportation routes throughout the metropolis (Figure 3) provide wildlife corridors between many of London's open spaces. The general pattern of railways is a radial one and so, too, is the older road system, which has been extended to include an orbital component. These corridors allow interaction between rural wildlife populations and those in the enclosed natural spaces.

This transportation net was built up over a long period commencing with the mud carriageways of the pre-industrial era. In addition there are many pedestrian and equestrian pathways in the suburban areas dating from the previous century.

Such corridors are useful in connection with the dispersal of animals, but they can also provide permanent habitat for some plant and animal species. For example, railway cuttings through chalk deposits are known to carry

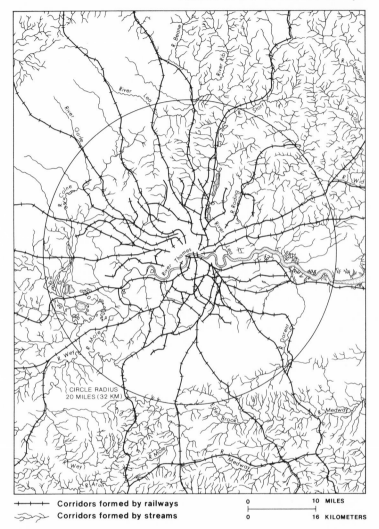

┼─┼─┼─ **Corridors formed by railways**
〜〜〜 **Corridors formed by streams**

Figure 3. Wildlife Corridors Formed by Surface Railways and Streams.

a diverse herbaceous cover and harbor a few plant rarities. They are therefore attractive to insects and can, where the soil is sandy, provide well-drained sites for fox dens and even rabbit warrens (Lond. Nat. 1963, 42:12; 1964, 43:135).

Undeveloped Land in the Metropolis

Many urban landowners develop only a small portion of their total acreage and much remains in a biologically productive natural state. Three major categories of urban land use may be considered in this connection: one may be termed institutional, another utilities, and the third domestic-commercial; these are illustrated in Table 8. It is often the case that institutional land abuts major open space and this is an extremely important factor helping to maintain public open space as useful natural habitat (Gooders 1965).

Table 8. Undeveloped Land in the Urban Area

Institutional	Utilities	Domestic-Commercial
Colleges and Universities	Sewage farms	Gardens (backyards)
Schools	Reservoirs	Market gardens (truck farms)
Hospitals	Canals	Industrial estates
Prisons	Gravel workings	
Nature reserves	Airports	
Clubs	Churchyards	
Corporations	Cemeteries	
Municipal offices	Rights-of-way	
Golf courses		
Race courses		
Zoological gardens		

One of the most productive public utilities is the sewage farm with its rich plant and animal life (Jones 1961; Gladwin 1963; Parr 1963). Technological change threatens to reduce the extent of this land use, with the old type of sewage farms being rapidly replaced with more compact, enclosed purifications works that are much less attractive to both resident and migratory species (Lond. Nat. 1966, 45:8).

VIABILITY OF SOME ANIMAL SPECIES IN LONDON

The fox and badger illustrate how different species have adjusted to the expansion of London into the rural land-scape. Teagle (1967, 1969) has written on the suburban ecology of both species and it is apparent that the fox has proved more successful in maintaining its status and invading the urban area, and has breeding locations within four miles of St. Paul's Cathedral. It is found in proximity to densely populated districts, and makes use of sheds and dumps to provide shelter for the rearing of cubs. It is indifferent to both traffic and noise, and manages to thrive under what at first might appear to be unpromising conditions. Despite the fact that the animal may suffer from the effects of air and water pollution it is less persecuted by man when living in the city than when in rural areas, and its life is therefore potentially longer.

By contrast, the badger usually loses the battle with the developers, being repeatedly forced to abandon its ancient dens (or "setts") and retreat to greener acres. One principal difference between the two animals is the greater control exercised upon the badger by the physiography of its environment. Unlike the fox, a badger normally excavates its own den and seems unable to live above ground permanently. It is, therefore, limited by the surface geology of the London region and cannot exploit the whole area. In common with the fox, the suburban badger suffers less direct harassment than his rural relatives because he is not seen as a major threat to crops and livestock. However, his massive excavations when conducted in the relatively small space of a suburban garden can try the patience of even the most ardent animal lover and this has been the principal cause for the badger's retreat to the open spaces and remnant woodlands.

This matter of environmental limitation will apply to some extent to all creatures that burrow. Surface deposits and soil types can control the distribution of moles, shrews, voles, rabbits, and birds such as sand martins or kingfishers (*Alcedo atthis*).

The problem of conflict with man is exemplified by the mole, which can cause tremendous losses in the flower and vegetable garden and wreak havoc to sports grounds and lawns. This insectivore spends it active hours burrowing through the soil in search of its favorite prey, the earthworm. Few gardeners or groundsmen are prepared to entertain such a visitor and the mole is trapped and gassed with great persistence. The result of the confrontation is the expulsion of the species from developed areas and its concentration in woodland copses, on commons, or in nature reserves.

This example can be contrasted with that of another insectivore, the hedgehog, which rarely comes into direct conflict with man. While man does not regard this animal with animosity, this species suffers great losses on the roads. Despite such high mortality, the records of the LNHS indicate that the hedgehog is probably the most widely distributed mammal in the London area. It is found in all the principal parks and thrives in the small gardens even where they are surrounded by high-density development. As long as wildlife corridors remain, allowing the species to move when renewal activity occurs and to recolonize the area once building activity ceases, this species will probably persist as an urban mammal.

The gray squirrel is another small mammal which has invaded London, and, indeed, is listed as a pest in England. It is possible that its arboreal habit has provided the key to success, and its ability to utilize a variety of food has further aided it.

Amphibians and reptiles are rare in many temperate zone cities. At least with respect to amphibians, this is to be expected in view of the loss of wetland and aquatic habitat; it has yet to be explained in what way artificial pools fall short of the requirements of these species. The common frog has become scarce in most urban areas and this may result from a combination of circumstances, as discussed in Chapter One.

Without the benefit of specific studies it may be premature to conclude that omnivores with flexible habitat requirements are the most prominent wild mammals found in the built-up districts of the urban area. However, it should be noted that the fox, hedgehog, and squirrel fall into this category.

FACTORS CONTROLLING THE DISTRIBUTION OF ANIMALS IN LONDON

LNHS distribution maps for mammals recorded in the London area reveal that only a very few species survive when restricted to the domestic garden zone. Most species are heavily dependent upon access to large open spaces with wild vegetation and the largest populations exist in the rural-urban fringe zone (Table 9).

Teagle (1967) illustrates that the distribution of wild land within this rural-urban zone is related to the kind of topography that prevails on the periphery of the metropolis. He comments: "The eastern Surrey fringe of London

Table 9. London Mammals

Mammals	Preferred Habitat
Mole (*Talpa europaea*)	Peripheral distribution
Stoat (*Mustela erminea*) Weasel (*Mustela nivalis*) Badger (*Meles meles*) Hare (*Lepus europaeus*) Rabbit (*Oryctolagus cuniculus*)	Urban-rural fringe
Mole (*Talpa europaea*) Fox (*Vulpes vulpes*) Squirrel (*Sciurus carolinensis*) Voles (*Microtus* and *Clethrionomys* spp.) Mice (*Mus musculus* and *Micromys minutus*)	Large open spaces throughout the built-up area
Hedgehog (*Erinaceus europaeus*) Shrews (*Sorex* spp.) Squirrel (*Sciurus carolinensis*)	Small gardens, hedgerows and parks throughout the built-up area

Compiled from: LNHS reports 1960—70.

represents a curious picture with glacier-like suburban flows filling long valleys between hills which have retained their woods and chalk grassland."

Figure 4 shows a high concentration of mammal observations in the area mentioned by Teagle with a noticeable decline in the eastern and northeastern part of the metropolitan fringe where the land is relatively flat and therefore almost completely built over. Thus maintaining the green belt seems necessary to retain wild populations. Mere preservation of existing land will not suffice. It will be necessary to practice ecologically based land management and take the opportunity of creating new open space and wildlife habitats when renewal schemes are first designed.

It is imperative that functional wildlife corridors in London be maintained or new ones created if the smaller green spaces are to remain capable of supporting stable populations. We do not yet understand to what extent

(Circle radius 20 miles)

Compiled from observations on 8 species; Badger, Common Shrew,
Fox, Hare, Hedgehog, Mole, Stoat and Weasel.

N? of species observed

□   1
▫   2
▣   3
▨   4
✳   5
■   6

*Figure 4.   Observations of Mammals in London Natural History
Society Area, 1956–1969. Shading indicates built-up area.*

urban wildlife populations interact with rural ones, or
whether there actually may be suburban areas that support
a greater number of breeding pairs, thus creating a surplus
population to migrate to more rural districts.

One further constraint which may yet limit the richness
of London's wildlife is environmental contamination, the
true extent of which has not yet been assessed. The fact
that the most suitable parks occur in the elevated and
therefore less polluted districts has already been pointed
out. However, these same parks are increasingly used to
provide short cuts for congested, rush-hour traffic which

pollutes the park areas with hazardous compounds. Parking of vehicles in central parks is a new practice and must have pronounced effects upon the quality of water seeping into roadside soils. Another perhaps less important problem is that of complaints from vehicle owners who find their cars bespattered with bird excreta or aphid "honeydew." Provided that this does not tend to removal of the offending fauna and trees, the use of these beautiful oases as daytime car dormitories may not be particularly detrimental.

Very evident from reports in the London area is the great disparity between the ability of birds and mammals to use the urban environment. The mobility of birds allows them to exploit ephemeral or temporary vegetation and they can survive by using a combination of non-contiguous green spaces that are not too widely separated. Such a strategy is not without danger, however, as was shown by Snow (1958) when he noted the high mortality suffered by blackbirds which fed in parkland across the road from a garden in which they nested. Many a nest was left untended because a heavily laden parent bird collided with a vehicle on its return journey across the highway.

Wildlife corridors are essential if mammals and other cursorial species are to be retained as components of stable urban ecosystems. The word *stable* is most important when considering enhancement of animal populations in cities. We have seen too many instances of unstable systems where one species reaches disproportionate numbers and is rightly deemed a pest. London provides a clear example of the correctness of the belief that stability is associated with diversity of habitat type, vegetation structure, and species composition. It is equally obvious that the essential requirement in cities is a well-planned management program to maintain habitat and create or modify it as required in the face of consumptive land uses.

LONDONERS—THE CULT OF NATURE

When discussing the factors which have influenced the apparent success of nature in the metropolis, it is necessary to consider the attitudes of its human residents. Fitter (1945) talked of the generally favorable attitude toward certain segments of nature witnessed among the residents of London, but indicated that their attitudes toward other living organisms tend to be ambivalent. Fowles (1970) believes that the English people possess a peculiar rapport with their surroundings and that this is exemplified by their passion for gardening and pet keeping. Many people are interested in natural history and are therefore less likely to designate all insects as "bugs" and then adopt a reflex response to spray anything which vaguely resembles a "bug." Clearly this is an interesting and important part of the subject, but it requires study on an objective basis. Whether it is true that contemporary Londoners are more tolerant of their wild neighbors is open to doubt; that they have by accident been handed a rich natural heritage of green space which happens to harbor a diversity of living organisms is demonstrably true.

In a review of the history of London's growth, Trent (1965) warned of the potential danger to the present Green Belt which, being an artificial rampart, can be demolished as easily as it was created. Already it has been breached in its weaker sections where development has trailed along roads and rail lines which cross the "green ring."

Trent's words suggest the potential danger to the non-human populations of the metropolis if their value is not appraised and objectives set for their management as an essential component of the rich and varied resource base of London.

CHAPTER FOUR

Los Angeles: A City with Islands of Wild Landscape

Dasmann (1966) feels that in the foreseeable future, the only wild animals that many of America's city dwellers can know will be those species that live within the urban environment. This chapter discusses how one animal, the coyote, has been retained in one of America's major cities, and how some of its residents perceive this urban carnivore.* The coyote persists within Los Angeles because historical accident and fortuitous topography have made its presence possible (Gill 1966; 1970). Its persistence within the city is not an ordinary situation, but this illustrates another facet of the ecological diversity that can be maintained in the urban landscape.

COYOTE HABITAT IN LOS ANGELES

The Los Angeles urban complex is a sprawling superimposition of man's technology upon an extremely varied

---

*Although classed as carnivores, coyotes in general and urban coyotes in particular have extremely varied diets. Similarly, insectivores such as hedgehogs, moles, and shrews (Chapter 3) do not subsist solely on insects. These terms, carnivore and insectivore, are taxonomic terms and do not necessarily describe accurately the diets of the animals to which they apply.

physical landscape. During the past few decades, most of the natural environment within this area has been modified or eliminated to an extraordinary extent, but much of the wild landscape remains.

Bulldozers have cut giant staircases into the more accessible acreages in the surrounding mountains to create sites for housing tracts. Yet thousands of acres of wildlife habitat remain on slopes too steep to develop, even near the heart of the Los Angeles metropolitan district. This habitat is comprised primarily of the chaparral association which is the most characteristic plant association for all but the desert areas of Southern California (Aschmann 1959). This community possesses a wide range of associated species, but the shrub chamise (*Adenostema fasciculatum*) is the most prevalent. The very dense and often thorny scrub vegetation functions as a substantial barrier

Dense chaparral covers bottom of canyons in Santa Monica Mountains adjacent to a suburban development. Coyotes are seen and heard in this area. In the background the densely populated San Fernando Valley is visible.

to travel by man, one of the reasons why remnant islands remain in a relatively undisturbed condition. Furthermore, this association occupies steep slopes, adding to its inaccessibility.

The coyote remains in many of these diminishing remnants of wild landscape within Los Angeles, giving testimony to its adaptability and hardiness. Young (1953) assesses the coyote's adoption of man-made environment with this statement: "The coyote has cast his lot with civilization. His fortress is barbed-wire fence; he can sleep under cover in a bed chamber walled by corn, wheat or grass; orchards and gardens are among his pantries." Ingles (1954) also illustrates the flexible nature of this animal: "The coyote has been severely persecuted with guns, traps, and poison; but in spite of all this, it seems to be holding its own in many places, a fact which speaks well for the adaptiveness of so large an animal." According to Dobie (1961), no other wild animal of historic times has shown itself so adaptable to change.

Aside from the coyote's adaptability, an important reason for the close proximity of this animal to an urban population is the abrupt transition that exists between the built environment and the natural habitat. A combination of very rapid urban growth (especially when compared to a city such as London) and restrictive topography have created a situation that is duplicated in only a few North American cities. Most metropolitan areas have transition zones between their urban and suburban, suburban and rural, and rural and natural or wild areas. Not so Los Angeles. Abruptly bounded on the west and south by the Pacific Ocean and on the north and east by mountains (Figure 5), Los Angeles is an urban complex that is sharply abutted by a wild landscape, with little opportunity for a border zone to develop between the two. Second, much of this wild landscape is linked to large areas of less disturbed habitat which creates corridors for wildlife movement.

This clearly aids the maintenance of an urban coyote population.

DISTRIBUTION OF THE COYOTE
IN LOS ANGELES

Coyotes live both within the city and in areas peripheral to it within the greater Los Angeles area. Some of their habitats are completely ringed by urban development and are becoming less important as reservoirs of wildlife populations. On the other hand, areas peripheral to the Los Angeles basin continue to maintain large numbers of the coyote.

Peripheral Distribution

The major peripheral coyote population is located in the abruptly rising Santa Susana and San Gabriel Mountains that form the northern boundary of the Los Angeles lowland. These mountains, which reach elevations of more than 10,000 feet, contain the Angeles National Forest, a large and primitive wilderness region which not only supports numerous coyotes and black bears (*Ursus americanus*), but also an occasional cougar. This important wilderness area contains the major population of coyotes in the Los Angeles region.

The coyote population of the San Gabriels is most often noted by urban man during the not infrequent times of ecologic stress, when wildlife is forced down the canyons into the adjacent suburbs. During periods of drought, fire, or diminished food or water supply, foothill residents increasingly report depredations by this predator. Accounts of complaints and sightings help to document the presence of coyotes in this area.

Internal Distributions

To the east of the Los Angeles lowland lies the densely populated San Gabriel Valley which acts as an effective barrier to wildlife movement. On the southeast, however, lie the Puente Hills, an outlier of the Santa Ana Mountains farther to the southeast. This region is rapidly being terraced and developed for housing and although the upper slopes are ringed by urban development the coyote still maintains small local populations. The westernmost coyote habitat in this area during the early 1960s was 11 miles east of downtown Los Angeles. In Whittier, along the western flank of the Puente Hills, a den containing ten coyote pups was discovered in a commercial area on the city's eastern edge (Froman 1961).

Until the mid-sixties, a relatively large number of coyotes inhabited Palos Verdes, a peninsula protruding into the Pacific some twenty miles south of downtown Los Angeles. Coyotes were gradually forced out by residential development until by 1965 only a few individuals remained.

The most important coyote habitat within Los Angeles is located in the Santa Monica Mountains. This portion of Southern California's transverse mountain range is comprised of many extremely dissected ridges that approach 2,000 feet in elevation. Here the most extensive tracts of chaparral habitat within the city are found. A 23-mile wedge of the Santa Monicas protrudes eastward into the heart of Los Angeles, dividing the city roughly in half. Nearly 40 square miles of deep winding canyons and narrow ridges continue to support a large reservoir of wildlife.

The largest concentration of coyotes in the urban portion of these mountains is along the canyons and ridges near the summit of the range. Much of this area also

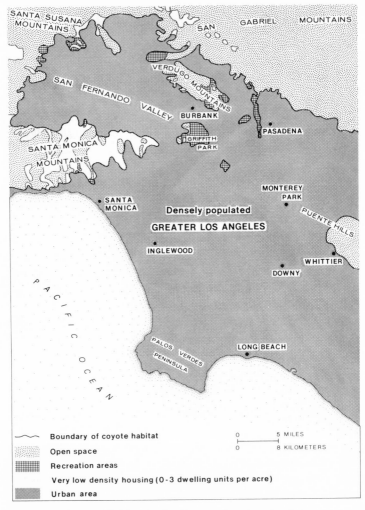

Figure 5. Coyote Habitat in the Greater Los Angeles Area.

provides sites for low-density, exclusive homes that occupy ridges above deeply dissected valleys. Although many of the ridges have been developed, most canyons remain densely vegetated, allowing wildlife to easily penetrate between the built-up areas. As a consequence, Santa Moni-

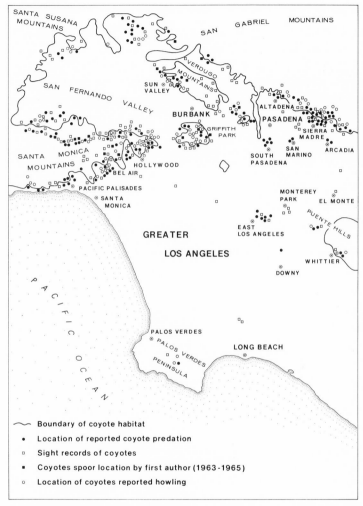

SANTA SUSANA MOUNTAINS

SAN GABRIEL MOUNTAINS

SAN FERNANDO VALLEY

VERDUGO MOUNTAINS

SUN VALLEY

BURBANK

GRIFFITH PARK

ALTADENA

PASADENA

SIERRA MADRE

ARCADIA

SAN MARINO

SANTA MONICA MOUNTAINS

HOLLYWOOD

BEL AIR

SOUTH PASADENA

PACIFIC PALISADES

SANTA MONICA

MONTEREY PARK

EL MONTE

PUENTE HILLS

EAST LOS ANGELES

GREATER

LOS ANGELES

WHITTIER

DOWNY

PACIFIC OCEAN

PALOS VERDES

PALOS VERDES PENINSULA

LONG BEACH

⌒ Boundary of coyote habitat

• Location of reported coyote predation

▫ Sight records of coyotes

■ Coyotes spoor location by first author (1963-1965)

○ Location of coyotes reported howling

Figure 6. Reports of Coyotes in the Central Los Angeles Area, 1955–1967.

ca residents have a more intimate acquaintance with the coyote than dwellers of more totally urbanized districts.

In ascertaining the degree of coyote presence in this area, however, it was necessary to critically evaluate the accounts of local residents. While many of those interviewed professed positive knowledge of coyotes ranging near their homes, several of their neighbors knew nothing of the animal. Some of those interviewed had only a vague concept as to what manner of animal the coyote was. Two of this group reported having seen animals in the chaparral near their homes that had the appearance of wild dogs, but they were unable to judge if they were coyotes. Others had heard repeated howling on various nights, but could not determine the cause of the sound. Another interviewee had recently lost a number of oriental fowl and felt they had been stolen or killed by "wild dogs" that she had seen in the nearby chaparral.

In January 1965, an investigation of various locations along Mulholland Drive, a street that follows the mountain summit, revealed substantial evidence that these semi-developed mountain tracts serve as coyote habitat (Gill 1966). Numerous sets of coyote tracks were found along canyon bottoms, and scats of varying ages were observed, indicating that coyotes maintained a permanent residency in this area. More recent evidence gained from the Los Angeles Department of Animal Regulation (LA DAR) (1971) indicates that in 1971 coyotes were still in abundance along Mulholland Drive though they were causing a minimum of problems with the local residents.

An unusual coyote habitat is located in the easternmost extent of the Santa Monica Mountains. This area includes Griffith Park. With about six square miles of ridges and valleys, it is reportedly the largest municipal park in the United States (Cunningham 1964). Wildlife and native vegetation are protected here, and since coyotes are not molested by any form of control, this portion of the Santa Monicas supports a relatively dense coyote population only six miles from the Los Angeles central business district. This anachronism has produced some rather unusual situations, such as interruptions of outdoor concerts

Chaparral provides dense plant cover, suitable for coyote habitat, around Hollywood Bowl. Concerts here have been interrupted by a chorus of coyote howling apparently triggered by the music.

by coyotes howling from surrounding ridges (Froman 1961), and coyotes preying upon caged animals in the Griffith Park Zoo (V. Wrineo, Night Animal Keeper, personal communication 1964).

The coyote in Los Angeles is not rigidly confined to areas of wild landscape. Occasionally it ranges into urbanized zones within the basin, as the following citations indicate: By 1937, the city's central business district was well established, yet a coyote was killed by an automobile while crossing one of its main streets (Los Angeles Times 1937). One year later, a coyote was shot near downtown Inglewood while preying on poultry (Los Angeles Times (1938a). A pair of coyotes was trapped in 1946 by a government hunter near Downey, nine miles southeast of downtown Los Angeles in an area already completely surrounded by paved streets and houses (Young and Jackson 1951). In 1950, police dispatched a coyote in a garage in West Los Angeles; this animal had previously been seen walking through a shopping district several miles to the west of the downtown area (Los Angeles Times 1950). Another coyote was killed during the mid-fifties in an urban area of Monterey Park, seven miles east of downtown Los Angeles (Los Angeles Times 1955). A notable incident within the city occurred in Long Beach, at the southern periphery of the Los Angeles lowland. A biology instructor at Long Beach State College reported that two coyote pups which had been caught on a nearby street were brought to him for identification. Since one pup was a month older than the other, it appeared that two separate coyote dens were located in an area where there was scarcely an acre not devoted to urban development (Froman 1961).

The distribution of the coyote is thus not restricted to those areas that retain a favorable wild landscape. Members of the species may range through all portions of the Los Angeles urban complex.

POPULATION OF THE COYOTE IN
LOS ANGELES

To establish the population size of any animal species is difficult, and to estimate coyote numbers in a habitat as complex as the Los Angeles area is an especially difficult endeavor. No enumeration of coyotes has been attempted in this region, but many persons have indicated that relatively large coyote populations have existed from time to time in some portions of the basin. In 1948 the San Francisco Chronicle commented that Los Angeles County contained the State's greatest population of coyotes: 77 coyotes were trapped or otherwise killed in and near Los Angeles during one month that year.

While such accounts suggest relatively large numbers of coyotes in Los Angeles, estimation of the total number proves difficult. As with most carnivores, the coyote increases, maintains, or decreases its population in proportion to the available food supply, which is an ecological element so variable in southern California that even approximate numbers are difficult to arrive at. Coyote numbers usually fluctuate in proportion to rodent and lagomorph (chiefly rabbit) population cycles. As a particular prey species begins to increase, coyote numbers rise proportionately. Due to disease or factors of ecologic stress, peak populations of lagomorphs or rodents are normally followed by a rapid decline, leaving coyotes with a lowered food supply. This in turn causes a decrease in coyote population, primarily through a reduction in the size of their litters.

If it were possible to predict rodent or rabbit population peaks, some correlation between rodent-lagomorph and coyote populations might be attempted. However, increase in rodent and lagomorph numbers in the Los Angeles basin is intimately associated with precipitation and consequent growth of vegetation, which is such an unpredictable factor in most semi-arid areas that a correlation between them is extremely difficult to draw (Thompson 1965).

Additional variables of diet also influence the Los Angeles coyote population. Since much of the subsistence of many coyotes originates from man's enterprises, variations in cultural activities can cause variations in food supply. Rotation of field crops, elimination of municipal dumps, laws concerning the ownership and restraint of domesticated animals, are but a few of these variables. Other elements of ecologic stress such as fires, floods, and drought which exert frequent influences upon the fauna of southern California, add to fluctuations in coyote population.

Predatory animal control by the LA DAR also causes coyote populations to fluctuate in size. When coyote numbers increase in a specific location, depredation to human property increases, and control attempts are focused upon the complaint area. The coyote population of a given locality is thus reduced, either through trapping or frightening the animals away. (Coyotes normally establish and maintain territories, but they will evacuate an area if sufficiently alarmed by control activities.) This latter type of reduction, while lowering concentrations in the complaint area, causes a population increase in adjacent locations, further complicating the distribution pattern.

The difficulty involved in estimating coyote populations in and peripheral to Los Angeles is thus clear. Nevertheless, certain criteria may be evaluated in arriving at a reasonable population estimate: (1) the number of complaints instigated by this animal as recorded by the LA DAR, (2) numbers of coyotes trapped or otherwise taken, (3) population estimates by persons concerned with the control of this animal.

Number of Complaints

From 1962 to 1971 the LA DAR received an average of some 115 complaints per year (Mason 1971). This many complaints during a nine-year period suggest a considerable population of coyotes in Los Angeles.

Numbers of Coyotes Taken

The County Board of Supervisors in 1937 began the first attempt at local control of coyotes. From November 1937 to June 1938, bounties were paid for some 650 coyotes (Los Angeles Times 1938b). Nearly 20 years later the number of coyotes taken during a similar length of time had hardly diminished.

The control program has continued in various forms. From 1961 to 1971 it accounted for a total of 349 coyote kills within the city (Weismann 1964; Mason 1971). During this same period, the U.S. Fish and Wildlife Service continued predator work outside the city limits in other areas of the Los Angeles basin. The number of coyotes taken by this agency has been surprisingly fewer than those trapped within the city—approximately 40 coyotes annually.

Combining the number of coyotes destroyed by trapping and other means of control, a total of more than 2,700 coyotes was taken in the vicinity of Los Angeles during the 10-year period 1961–71, an average of nearly 280 per year.

Population Estimates

To maintain yearly more than 100 complaints and to annually take some 280 animals suggests a large and permanent local population. Estimated coyote numbers within the restricted area of Griffith Park help to bear out this supposition. During 1965, animal keepers there estimated that a minimum population of 20 coyotes inhabited more than five square miles of park property. This number was also agreed upon by the Griffith Park Zoo Director (Young 1965). One of the most insular areas of wild landscape within Los Angeles thus supports nearly four coyotes per square mile. This estimate cannot be applied to other areas inhabited by the species but it gives an indication of the coyote numbers which the wild landscape of Los Angeles *can* support.

Various informal evaluations of coyotes in Los Angeles by knowledgable personnel put the number at several hundred in the city and several thousand in nearby peripheral areas. On the basis of our own investigations which showed a significant quantity of coyote spoor during two years of study through most of the Los Angeles wild landscape (Figure 6), we conclude that during the past 10 years a minimum population of 400 coyotes has lived within the city, and an additional number of at least 2,000 has inhabited the mountainous peripheries of the Los Angeles basin.

REACTION OF THE COYOTE TO THE BUILT ENVIRONMENT

The coyote's most significant reaction to man's proximity has been adaptation; the fact that considerable numbers of coyotes remain in urban areas rather than emigrating to regions of less human pressure bespeaks this adaptability.

Many examples of adaptation can be cited: When the Hollywood Freeway was under construction during 1954, empty houses in the path of the freeway awaiting destruction had coyotes living under them (Mason 1965).

The way in which the coyote has adapted its behavior to take advantage of the wider food base offered by an urban environment is another striking example of its adjustment. Coyotes are frequently observed stripping fruit from the lower branches of the trees in orange, avocado, or plum orchards. They also feed upon such truck-garden crops as watermelon, cantaloupe, and celery (Froman 1961). Grapes and dates appear to be a favored seasonal food; Jaeger (1950) cites evidence that during certain times of the year some coyote feces are almost totally composed of date seeds.

As attested by the numerous complaints of pet owners in Los Angeles, the coyote as predator has found a plentiful and easily obtained source of animal protein in immature dogs and adults of smaller breeds as well as in the domestic and feral housecat (*Felis domestica*) (Los Angeles Times 1960a, 1960b; Mason 1965; Pasadena Star-News, February 7, 1970a; LA DAR 1971). In the course of our investigation an analysis of coyote feces from the Brentwood area of Los Angeles showed that cat hair was present in three separate coyote scats. Coyotes have also

At night coyotes commonly use storm drainageways, such as this one in Los Angeles, as routes into the city from adjacent undeveloped areas. Fences along the sides of such routes fail to keep the animals out.

been recorded preying upon animals maintained as live props on movie studio lots (Froman 1961).

Coyotes in Los Angeles are able urban scavengers: they even forage quite frequently from garbage cans (Weismann 1964). A greater indication of the adaptability of some coyotes to urban man is their acceptance of food purposely fed to them by individuals. One interviewee stated that food which her family places for coyotes in their back yard is regularly taken by the animals (Salzman 1964). Another woman routinely feeds candy to coyotes (Weismann 1964). One Pasadena resident became locally "famous" for feeding wild animals, especially coyotes, in her back yard (Pasadena Star-News 1970b). While these latter citations should not be construed to apply to the general coyote population of Los Angeles, they nevertheless indicate the extent to which some have changed their behavior.

An observation by the Griffith Park Zoo Director may indicate one result of this animal's adoption of an urbanized environment. Because of the broader food base and the ease with which food is procured, Young (1965) states that the coyote within Los Angeles is a larger animal than the species in the wild, a strong indication of favorable response to a man-modified environment. The coyote has proven to be very flexible, and perhaps the species reaches its greatest adaptation to man in the Los Angeles urban landscape.

COYOTE AND URBAN MAN: CONFLICT

Urban people do not universally understand wild animals and do not always appreciate them (Kirkpatrick 1971). Lack of appreciation of the coyote by some Los Angelenos has inevitably led to conflict between the animal and man. The coyote "problem" in Los Angeles is not a recent development; as early as the 1930s newspapers carried accounts of coyote-caused damage (Los Angeles Times 1936; 1938b). Further complaints caused by forays of the coyote continued to be numerous in the 1940s, including reports of depredations upon domestic animals very near the City Hall (Los Angeles Times 1943).

Not only by predation has the coyote caused numerous protestations. This animal often exhibits a remarkable and well-known ability to howl, in a pitch and at a volume quite unacceptable to some residents. Since this primarily nocturnal mammal has the tendency to howl at an hour when most persons are attempting sleep, it is understandable that complaints may arise from this type of disturbance. Several accounts indicate that coyotes howl* nearly every night through back canyons of heavily populated districts.

One area of great potential conflict that has never developed into a problem in Los Angeles is that of the coyote as a transmitter of rabies. According to K. H. Sutherland, Los Angeles County Health Officer, until 1965 the coyote had never figured in a rabies case in the county (personal communication). From 1965 to 1971 the coyote continued to pose little threat as a vector of rabies (LA DAR 1971). This is notable in view of the coyote's role as a carrier in rabies outbreaks in other areas, and is more remarkable when it is realized that skunks, which have been a major rabies problem at times within Los Angeles (Mason 1965), are preyed upon by the coyote; it seems surprising that the

---

*An incident that occurred in Los Angeles during the Mexican War (1846–1848) gives a humorous illustration of the magnitude of the coyote's howl. One of the sailors ("Webfoot," 1871) under the command of Commodore Stockton wrote:

"In the night at Temple's Ranche, an alarm was given. The enemy was approaching with yells and shouts. The long-roll sounded, and men sprang to their arms. They were thrown into position to repel a charge, and momentarily expected General Castro's army to appear. The cries and yells continued but no attack was made; they remained under arms for about two hours, and would probably have kept so until daylight, had it not been for an old Indian. He informed us that our alarm proceeded from a couple of coyotes. . . . The hideous noise coyotes keep up resembles (in combination) the howling and yelling of various animals, the shrieks of women, the crying of children, and the barking of dogs. . . . While these two insignificant animals were keeping over three hundred men under arms for two hours, General Castro fled from Los Angeles."

virus is not know to have been transferred to coyotes here. Nor is the coyote considered to be a serious threat as a rabies vector elsewhere in the state. Between 1960 and 1969 there were only eighteen reported cases in California of coyotes with this disease (LA DAR 1971).

MANAGEMENT OF THE COYOTE

The result of conflict between *Homo sapiens* and any other animal invariably leads to the former's attempt to subjugate the latter. Endeavors to control the coyote probably have stirred more controversy in North America than efforts to control any other animal. Certainly more effort has been expended and more money has been spent on the control of the coyote than on any other predator (Dobie 1961).

One of the first official attempts at predatory animal control in Los Angeles County was made in 1937, when a bounty was levied against the coyote, but this type of control was considered ineffective. There followed several other attempts to devise effective control programs, and these were ardently supported by some citizens, and just as vehemently opposed by others.

In November, 1960, rabies broke out in the skunk population of the northern portion of Los Angeles, and in the ensuing nine months 36 skunks were proven positive rabies carriers. Extreme concern developed in the residential areas where rabid animals were located. It was especially feared that rabies would be transmitted to the larger carnivores, particularly the coyote (Mason 1965).

By January 1961, fear of a rabies epidemic lead to the initiation of an intensive program of wildlife control within the city. Money was appropriated by the City Council to hire contract trappers, in addition to those already on the payroll to control skunks and other animals in rabid areas. At the same time, final steps were being taken to authorize the Department of Animal Regulations to establish an effective predatory animal control program with the primary purpose of containing the coyote.

Contrary to the wishes of local conservation groups, an ordinance was finally passed which gave Animal Control Officers legal authority to destroy predatory animals. The control program called for the elimination of the individual animal which caused the actual complaint; its purpose was not to extirpate the coyote from Los Angeles but to provide a controlling drain on the population, especially eliminating those individuals that perpetrated specific damage. For the ten-year period from the beginning of the control program to 1971, a total of 349 coyotes was trapped within the city (Weismann 1964; Mason 1971).

Other management methods have been recently initiated: in September 1969, what was termed "a unique wildlife conservation program" was begun by the LA DAR. This program, named "Animal Airlift," takes humanely trapped urban wildlife (including coyotes) and transports them by helicopter to the Angeles National Forest 35 miles northeast of the city. Between September 1969 and December 1970, a total of 712 wild animals, birds and nonpoisonous reptiles were relocated in weekly helicopter flights (Bryant 1970). The ground release sites are selected by the Game Manager of the Angeles National Forest, who locates areas where food, water and shelter are available.

While the Department of Animal Regulation is clearly enthusiastic about the airlift plan, the scheme does pose some problems. As of January 1971, no research had been conducted to determine whether the aims of this plan were being achieved (Bryant 1971). For example, tagging of released animals had not been established to determine whether wide-ranging animals such as the coyote were returning to the city. Second, if the localities where animals are released are already at their carrying capacity, the survival rate of these immigrants may be poor, particularly since most of them are adapted to a suburban habitat where the food base is quite different from the diet available in the Angeles Forest.

The relocation plan may be basically sound, but research must be instituted to determine if it is achieving its aims before it can be called a significant conservation tool. It represents, in any case, innovative thinking—something that is surely necessary in any program of urban wildlife conservation.

Despite a variety of control efforts by the LA DAR, coyotes within Los Angeles remain relatively protected. The city does not use poison such as Compound 1080 to control pests (LA DAR 1971), and firearms or traps may not be used by anyone other than Control Officers. Since much of the coyote's habitat is on private property that is fenced against trespass, it enjoys added protection. The coyote in Los Angeles thus has few enemies, except for the limited number of officers who make a career of controlling the individual animals that create complaints. Since there is no great check on coyote numbers, a continuing population within and peripheral to the city seems assured.

THE COYOTE AS AN ESTHETIC RESOURCE

In the process of creating today's technological civilization, contemporary man has become more and more divorced from the land. The natural landscape and its wildlife are no longer the familiar objects they were in a once rural North America. We are beginning to reassess the value of the natural landscape, of which animals form a most vital component. Nature has begun to assume a greater significance, particularly to urban man.

The coyote is one element of the wild landscape which in the past has often been looked upon with disfavor; but social values change. As our culture becomes more urbanized, a view is developing that wildlife, including the coyote, is a resource that deserves preservation rather than persecution.

A man's personal opinions concerning the worth of a particular resource are variable and subjective, thus difficult to measure. However, general response patterns are observable and can be assessed. When varying opinions are directed toward an animal resource so controversial as the coyote, they become more measurable. Controversy solidifies opinion; convictions can be evaluated.

The controversy that developed in Los Angeles during 1960 and 1961, when the Department of Animal Regulation sought legislation to permit the trapping of coyotes, is one example of opinions that may be assessed. Due to predation of pets and other property, and the fear of rabies, many persons wanted to rid the Los Angeles area of this animal; yet a surprising number protested any molestation of the coyote. Many claimed in defense of the animal that they had purposely moved to hillside areas to be nearer wildlife (Los Angeles Times 1960a). During this period the City Commission received approximately sixty complaints relating to the presence of coyotes in the Hollywood Hills area, but also received a petition bearing sixty signatures asking that the coyote not be molested (Los Angeles Times 1960b). The most significant of the positive reactions toward the coyote was an appreciation for the feeling of remoteness from the city which the coyote evoked when seen or heard.

Many persons interviewed during this study were of the opinion that our society is moving ahead at too rapid a pace, and the presence of the coyote near their homes gave them a satisfying sense of intimacy with nature. Others expressed the thought that the coyote is a picturesque animal that reminded them of their American pioneer background.

In 1965, as a result of proposed liberalization of the City Trapping Ordinance, a member of the Board of Directors of the city of Pasadena (near Los Angeles) received more than one hundred letters in defense of the coyote. His summary of the more eloquent pleas revealed the attitude of many urban dwellers toward wildlife: "The coyote population in the local hills is part of the colorful heritage of the West. It should be considered a privilege to live next to wild areas and wild animals. Children should benefit from such an environment" (Pasadena Star-News 1965).

A statement by the Secretary of the Interior's Advisory Board on Wildlife Management documents the esthetic value of the coyote on a statewide basis. Their report (Leopold 1964) testifies that in many areas of California the esthetic value of the coyote greatly exceeds any potential damage it might cause.

As America continues the rapid transformation of its landscape from a natural one to a cultural one, less and less remains of the original ecosystems. Since the coyote is capable of maintaining its range in a man-altered urban habitat, it may eventually become one of the few large wild animals remaining in all but the more remote sections of the rural environment. Thus any negative economic value of the coyote's depredations on man's property will be balanced by its positive value as an esthetic resource. When wild nature is an integral part of the city ecosystem the esthetic awareness that is engendered in urban man can become a vital force in enhancing the quality of urban life.

CHAPTER FIVE

Planning for Wildlife in the City

This chapter is principally concerned with North America because we feel that the greatest opportunities for urban wildlife management and creation of natural areas exist on this continent. In western Europe, nature has been retained in cities to a much greater extent, and while opportunities for urban wildlife management exist, the need to exploit them is not as great. Also, the pace of development is much accelerated in North America, and therefore the urgency of the problem is significantly greater.

The examples of planned open space or natural areas cited here are drawn from both Britain and the United States, but there is a noticeable discrepancy between the two countries. In the United States and Canada there has been relatively little research on non-game species, whereas in Britain, since the establishment of the Nature Conservancy in 1949, much knowledge has accumulated concerning the management of a great variety of animals. Added to this is the expertise of a large body of amateur naturalists who, through their County Naturalists' Trusts, have accepted responsibility for the management of many nature reserves, areas of special scientific interest, and municipal commons.

OPEN SPACE IN THE URBAN REGION

In North America open space has been the subject of much attention in recent years, principally as a consequence of its inequitable distribution in most cities. The contemporary pattern (and general paucity) of undeveloped land in the urban matrix has resulted from the economic pressures which typify our free enterprise society. Fortunately a growing literature describes the various methods by which urban land can be secured for public, recreational, and amenity purposes (Clawson 1962; Mandelker 1962; Whyte 1962, 1968; Strong 1965; Little 1968; Brooks 1970; Kent 1970; Shomon 1971; Smith, N. C. n.d.).

Open space—land and water in the urban region not covered by artifacts together with the space and light above them—has a three-fold function (Tankel 1963). It has beneficial psychological effects upon those who enjoy access to it, and it helps shape urban development. Ironically, the third and apparently most important function has been the least appreciated. That is the capacity of open space to protect natural landscape features which are essential contributors to the ecological stability of any region (Goodman and Freund 1968). Concern for urban open space has been heavily oriented toward its active recreational function with little or no recognition of amenity aspects (Mann 1970).

Open Space Planning

Traditionally open space planning has utilized arbitrary space standards for recreational activities based upon population density, and it often bears little relation to the real needs of the urban populace. Inevitably there has been a continuing loss of green space under bricks, asphalt and concrete, despite the lowering of environmental quality

To control erosion, banks of this stream in a park in Baltimore, Maryland, have been protected with concrete. Sometimes this may be advisable, but such streams then provide virtually no food or cover for wildlife, and add little, if anything, to the diversity of flora or fauna of the area. In addition, such concrete borders reduce the esthetic appeal of streams. (*Silvan photo.*)

which this entails (Tunnard and Pushkarev 1963; Dickinson 1966; Clawson 1969). In Britain the green belt concept implemented in the 1940s has largely succeeded in counteracting the inexorable sprawl from the city center but it is a restrictive rather than a constructive tool.

A Holistic Approach to Land-Use Planning

Newly emerging planning concepts may mean that the day is fast approaching when open space will be accorded high priority in land-use conflict (Farness 1966; Hansen

1968; Zisman *et al.* 1968). A growing number of planners display an awareness of the need to plan all resources as part of an interdependent system. The principle that there is a "best" use for any given piece of land is now seen as the fundamental determinant of land-use planning. The notion that the equitable distribution of resources can be satisfactorily brought about by the allocating mechanisms of a monetary-based market economy is at last being cast aside (Niering 1960, 1968; Whyte 1962, 1964, 1968; Tankel 1963; Downes 1965; McHarg 1966, 1969; Dasmann 1966; Benthem 1968; Hackett 1969; Gottman 1970).

Acceptance of this principle implies a regional or a national perspective and ultimately a global approach to resource planning, since most contemporary boundaries have been fixed without benefit of these ideas. It also requires that the artificial division between so-called urban and rural planning be removed. A systems approach, based upon a desire to see the whole rather than the parts, is necessary if we are to attempt to understand and seek solutions to pressing contemporary problems (Caldwell 1966; Woodbury 1966; Davis, A. A. 1967).

Open Space—The Future

One champion of open space in the city (Wingo 1963) foresees an increase in the amount of urban undeveloped land in North America due to the greatly dispersed pattern of urban settlement which can be expected in the future. In his words the pattern will comprise "a loosely knit, weakly centred, low-density urban region spread over a wide hinterland." If such a trend is realized, it will be essential that it be accompanied by appropriate planning to maintain environmental quality. Already on the North American and European continents conventional, rigorously applied low-density standards are disappearing in favor of the idea of totally planned communities where clustering allows high density building in conjunction with extensive open land as an integral part of the community. Where it is possible to incorporate cluster development into regional plans so that all green space becomes linked by natural corridors, the total usefulness of the land is greatly enhanced.

The attractiveness of such rational planning is emphasized by Zisman et al. (1968) who point out that at our present rate of urbanization 30 acres accomodate 100 people, but that in new, rationally planned communities utilizing clustering and having abundant open space, only 10 acres per 100 people are needed.

In addition to the evident amenity and recreational potential of planned open space networks, another function has been recently highlighted. What might be termed an ecological rationale has been proposed by Odum (1969) and a complementary technological reasoning is advanced by Spilhaus (1971). Their statements focus upon the desirability of creating urban areas interspersed with open space linked to the peripheral countryside. Odum proposes a landscape compartmentalized in such a way that a spatial intermixing of variable-age ecosystems is achieved, thereby providing an overall stability. Such a landscape would fulfill the requirements specified by Spilhaus in order to bring about closure of the many materials cycles whose residuals currently create pollution problems. There is an agreement that we need a coupling of the town and countryside, an agro-industrial symbiosis, to efficiently utilize and conserve both human and natural resources (Dickinson 1966).

OPPORTUNITIES FOR PLANNED
WILDLIFE AREAS

Shomon (1970) advocates that attention be paid to the provision of "open living land" in the city, where wildlife would naturally be found. He and Dasmann (1966) both express the opinion that a city capable of supporting both human and numerous other animal populations will be eminently more liveable than one which attracts nothing more than sparrows, rats, and pigeons.

If the predictions outlined in the previous section are realized then it is clear that an increasing number of opportunities will emerge for the retention of natural areas prior to development, both inside the city and in the hinterland.

OPPORTUNITIES FOR ESTABLISHING
URBAN WILDLIFE HABITATS

Urban Renewal

Such schemes commonly involve either the removal of a number of small properties with relatively little attached land or the demolition of a large residence in spacious grounds. In each case it is possible to create excellent songbird habitat, an achievement entirely consonant with creating improved housing for people. Unfortunately the isolation of these areas from other green areas often limits their usefulness but the judicious planting of cover and food vegetation will allow some colonization. Fosberg (1966) believes that an attempt should be made to restore conditions resembling those which existed prior to the original settlement, but this idea may be unrealistic.

Planned Communities

In North America and Britain the completely planned town is becoming commonplace, and with it lies a great opportunity to establish integrated natural areas adjacent to all buildings (Wilmers 1971). Whyte (1968) has described in detail the concept of cluster development and the scope which this affords for maintaining areas such as river valleys, marsh, and woodlands.

Industrial Estates

With an increasing emphasis upon service industries and light engineering trades, many industrial estates are essentially similar in physical form to suburban residential areas, and they could equally well support songbird and small mammal populations, if appropriate cover were established.

Institutional Grounds

Many such grounds already offer considerable protection to wildlife, since they are essentially parkland. University campuses have become more dispersed in layout and often present a truly pastoral image. Their use on a seasonal basis especially aids the maintenance of those wildlife species that breed in late spring and thus experience relatively undisturbed conditions. Hospital grounds are even better in this respect since the degree of human intrusion is minor.

Schools have rarely provided very useful wildlife areas due to the small acreage involved, their heavy usage, and a lack of understory vegetation. Close-cut lawns and mature trees dominate such grounds, if, indeed, any vegetation is present at all. However, in view of the potential value of "natural" school grounds for developing a consciousness of the environment and a sensitivity to it, and for specific environmental studies, increasing the biological diversity of school grounds should be encouraged.

Zoological and botanical gardens have always harbored wild species in addition to those forming their collections. It is possible to increase the natural flora and fauna but care is needed to ensure against disadvantages such as the transmittal of exotic propagules and pathogens.

Corporate headquarters in the United States and Canada are increasingly being removed from downtown locations in major cities and relocated in suburban developments which offer an opportunity for imaginative landscaping and conservation of wildlife. One good example of this is the new American Can Company headquarters in Greenwich, Connecticut. (The social and economic issues involved in such corporate moves to the suburbs are important ones, but are outside the scope of this book, though it seems appropriate to mention that corporations might well be encouraged to improve the city landscape instead of moving away.)

City Parks

Traditionally, city parks have been formal in design with an emphasis upon decorative flower displays, close-cropped lawns, and generally antiseptic appearance. However, the "manicure syndrome" is disappearing, partly as a result of the recent steep rise in the cost of labor. The decrease in maintenance staff has to some extent been offset by mechanization, but the likelihood of additional park land reverting to a natural state has grown considerably in the past several years. There is thus scope for imaginative, ecological planning to produce safe but interesting wild areas which require a minimun of maintenance.

View of Lake Roland in Baltimore, Maryland, shows how part of the shoreline is maintained for people's use (note walkway and mowed grass), though much of the lakeside is untended and provides a suitable environment for many kinds of wildlife. No road approaches most of the lakeside, and this helps limit the number of people entering the natural areas. (*Silvan photo.*)

Derelict Land

Urban space that remains from the operation of extractive industries (for example a gravel pit) or is left following demolition of buildings is derelict land (Civic Trust, n.d.; Jensen 1967; Collins 1970). A report issued in England in 1963 estimated that some 100,000 acres located adjacent to urban centers were suited to reclamation for development or amenity use (Min. H. L. G. 1963). This type of land is usually associated with older cities and offers a real challenge to the resource planner. Attempts to restore such sites have too often resulted in a "cosmetic" operation producing flat, green expanses similar in appearance to agricultural land. To restore such land by modifying only those factors that inhibit the process of natural re-vegetation, and by retaining topographic irregularities so that a variety of micro-habitats exist, would be better.

If we retain the natural unevenness of the surface, we increase the probability that a variety of different soil moisture and incident light conditions will occur; therefore a wider range of plant species may find suitable growing conditions in the reclaimed area. Recent trends in the disposal of municipal garbage where organic material is separated and processed, eventually to yield acceptable compost, could make reclamation of derelict urban land economically feasible (Hart 1968; Carlson-Menzies 1971; Meyer 1972).

Another form of surplus land which has excellent potential for conversion to large areas of wildlife habitat is abandoned farmland. As a result of land speculation it is frequently unprofitable for farmers to continue operations once a city extends to neighboring property. (The mechanisms of this process differ between North America and Britain, but the results are similar.) A considerable acreage may come into this category and some thought should be devoted to securing strategically located parcels and protecting them from suburban sprawl. Such land could ideally be turned to use as educational nature reserves where the process of secondary plant succession could be witnessed at first hand. These acres also could provide excellent sites upon which to conduct planned experiments in habitat creation.

PRINCIPLES OF ESTABLISHING NATURAL AREAS FOR WILDLIFE

Certain environmental demands must be met when choosing land for the propagation of wildlife. Three requirements in any habitat are cover, food, and water in close proximity.

Inevitably, when dealing with urban areas, land rarely will be designated as single use; rather it will be expected to perform a recreational and protective function as well as to harbor wildlife. Under such conditions, however, key areas should also be established and managed primarily to benefit natural wildlife populations in order that they may function as regional reservoirs.

Distribution

Location of natural areas in urban centers should be based upon an evaluation of the adjacent land uses. Steps should be taken to ensure that external forces such as pollution do not adversely affect the natural occupants. If, of course, the vegetation is being used to reduce particulate pollution or buffer excessive noise, then its location will be predetermined (Leonard and Parr, 1970; W. H. Smith 1970).

To conserve urban wildlife resources, it is essential to maintain the physical continuity between open space units by such features as river valleys, hedgerows, or footpaths. The number and spatial distribution of vegetated areas will affect their ability to support animal populations. The network of undeveloped land in the urban matrix may be viewed as a system of interacting centers of activity where a change in the size or function of one will influence others. For example, an alteration in the type or amount of human use of a natural area will have significant effects upon the flora and fauna, and may cause the sudden dispersal of animals to other centers. Similarly, a reduction in park size by loss of land to development may result in its becoming nonfunctional as wildlife habitat, thus necessitating migration of its wild occupants. It is also evident that any change in the potential for interaction between centers will bring about a concomitant alteration in the viability of those centers. For instance, valuable city parkland located in a river valley could be threatened by highway development. The loss of such a corridor could significantly influence local animal populations if the valley acted as a means of introducing individuals from rural areas or if it acted as a reservoir of breeding animals which dispersed to nearby gardens and parks.

Size

We know that adequate size of any habitat is necessary to support a self-perpetuating ecosystem. Unfortunately we rarely can be certain what minimum acreage is required for any general habitat type. The relief, drainage characteristics, climate, soil parent material, and vegetation cover of a specific landscape unit determine its carrying capacity for wildlife populations. In addition, the past and present human use of the land greatly affect this carrying capacity.

Where feasible, the boundaries should be delineated with reference to known facts about community interaction and territorial requirements of the animals. Most species depend upon access to two or more habitats during the course of the year, and this should be recognized when setting boundaries.

Minimum sizes have been suggested for different habitat types but little data exist to substantiate them and the results of studies often conflict. For example, waterfowl populations vary in number according to the ratio of perimeter to surface area of water bodies, and Miller (1969) believes that the most desirable ponds are from one to five acres in extent. Atkinson-Willes (1969) on the other hand states that 100–150 acres is the size which produces the greatest waterfowl density.

Even very small waterbodies can be useful as habitat, provided they are grouped together. Such clustering is applicable to most forms of habitat and should be an important consideration in urban areas where the size of individual landscape units will almost certainly be restricted.

For songbird habitat it may be feasible to manage as little as one-tenth acre if it is planted with a mixture of mature trees, shrubs, and understory vegetation (Leefe 1968) and if it is near other areas of suitable habitat. This small size will be inadequate, however, if isolated plots are established. Williamson (1967) suggests that vegetated "islands" of one acre are the minimum viable-sized units of woodland.

Species present only during a part of the year, such as wintering species not present in the summer, may be important and interesting parts of a city's wildlife population. The area that such a "visiting" species requires may differ from that needed by a species that lives permanently in the city, a point that is important in planning or managing urban natural areas.

Another important consideration is that the more mature the ecosystem, the larger the area it will need to be ecologically stable. It is possible to create a small grassland habitat but not an equally small forest, since the "edge" effect—the tendency for increased variety and density where different natural communities join each other—will essentially preclude the establishment of true forest conditions. According to Moore (1962), the smallest size for any habitat is that which enables its essential species to maintain a stable population.

Ecological Principles

In the planning and creation of urban wildlife habitat certain principles should be followed. The underlying assumption is that genetic and structural variety in the natural environment is both desirable and necessary (Westhoff 1970). In Chapter One we suggested that diversity in ecosystems is associated with stability; if stability is desirable then diversity becomes a paramount objective in designing urban wildlife habitat (Leopold 1933; Pimlott

1969). Four manifestations of diversity are recognizable in any ecosystem, namely, stratification and niche differentiation, biochemical diversity, a range of species types, and the relative abundance of species (Odum 1969). The land selected for urban wildlife sites should be capable of supporting a community comprised of indigenous species grouped in natural patterns to provide maximum stratification for that community. Topographic variation in the landscape adds to the biological richness because of the several habitats created by the varied micro-environments.

Very closely connected to the concept of maximizing diversity is the idea of establishing a self-renewing community. If biological richness is achieved, then it is highly probable that a mature ecosystem will develop, provided that external perturbing forces (such as excessive physical disruption, intrusion of toxic substances, or changes in the level of the water table) are controlled to guard against regression of the system. Introduction of alien species should be avoided wherever possible.

Habitat Creation

Depending on the objectives and the time span involved, there are two major approaches to creating or restoring habitat. First, conditions may be established which are suited to natural colonization, and normal successional processes—as when the edges of a marsh fill in, become grassland, then shrub land, and finally forest—over perhaps a century or more will provide cover. Second, a higher degree of manipulation may be used, as when semi-mature vegetation is planted to speed the onset of shrub and tree cover. In an area with extreme environmental conditions, such as those inducing high rates of water loss through evaporation, the latter approach could be unsuccessful. A detailed knowledge of the local environment is necessary to manipulate conditions successfully.

One essential component of animal habitat is water, which is very often missing or is present in insufficient quantity in cities. The creation of small pools, streams, canals, or marshes could make an otherwise inadequate urban habitat attractive to small mammals and birds.

Linear habitat such as hedgerows, uncemented stone walls, a green area bordering a road, railway cuts and fills, canal banks, stream sides, pond margins, and utility rights-of-way can promote biological richness through its capability of providing shelter, food, and nesting areas for numerous animals (Odum 1959; Graham 1947; Johnston and Odum 1956; Niering 1968; Appleton 1970; Hooper and Holdgate 1970). Ecological studies of abandoned farmland can also provide useful information for wildlife managers in the rural-urban fringe; those of Beckwith (1941, 1950), Marks (1942), Spooner (1942), and Minckler (1946) are recommended.

There is a considerable amount of useful information—more than it is feasible to include here—about restoration techniques for derelict land and the subsequent recolonization of the land by plants and wildlife. Helpful references include Croxton (1928), James (1939), Yeager (1941), Tyner *et al.* (1948), Whyte and Sisam (1949), Limstrom (1953), Heyl (1954), Verts (1957), Holliday *et al.* (1958), Holliday (1961), Funk (1962), Koller and Knabe (1962), Coates (1964), Knabe (1965), Weston *et al.* (1965), Augustine (1966), Crompton (1966), Karr (1968), Antonovics and Bradshaw (1970), Collins (1970), Peters (1970), and Shetron and Duffek (1970).

Available evidence indicates that we can create in a city a range of habitats representing all the successional phases if we have a wide choice of sites, adequate funding, and access to local ecological data. A warning note must be struck in view of the scarcity of the latter: it is essential that a very careful appraisal be made of the most appropriate natural system for a particular city. Particular habi-

tat types exist in various urban areas and these specific
cases should be studied before attempting to create similar
features in other cities (Table 10).

The first essential step in any such planning is to conduct an ecological inventory before any manipulation is
commenced. Fosberg (1966) warned that a major problem
to be expected is the difficulty of knowing whether a

When land along power lines (left) and railroads (above) remains undeveloped it provides linear habitat which may connect a network of interacting nodes of wildlife populations. (*Silvan photos.*)

complete, viable community has indeed been established. One inconspicuous species such as a pollinating insect may determine the vital activities of a dominant plant and its absence could therefore mean short-lived success for the community. The basic guideline in all habitat creation is a cautious approach. Small-scale experiments, during development of a program to establish natural areas, may help to avoid serious large-scale problems.

Table 10. Cities Known to Have Significant Wildlife
Areas within Their Environs.

Location	Natural Area	Size
Amsterdam, The Netherlands	Woodland—*Amsterdamse Bos* (includes lake, marshes, grassland and woods)	2,200 acres
Oostelijk Vlevoland, The Netherlands	Lakes Wooded areas (bird reserves of grassland, marsh and woodland)	11,000 acres 7,500 acres
Dublin, Ireland	Phoenix Park (parkland with fallow deer herd)	1,760 acres
	North Bull Island (waterfowl and water reserve of saltmarsh, sand dunes and mudflats)	
Craigavon New Town, Northern Ireland	Planned retention of important biological sites and creation of wildlife habitat	
Frankfurt, West Germany	Enkheimer Ried (wetland bird sanctuary)	
Hamburg, West Germany	Fischbeker Heide (woodland nature reserve)	
Ann Arbor, Michigan, U.S.A.	Island Park (waterfowl area)	
Chicago, Illinois, U.S.A.	Cook County Forest Preserve (woodland)	40,000 acres
Milwaukee, Wisconsin, U.S.A.	Juneau Park (waterfowl lake)	
New York, U.S.A.	Jamaica Bay Refuge (Long Island) (tidal marsh system)	11,840 acres
Philadelphia, Pa., U.S.A.	Tinicum Wildlife Refuge (marsh associated with tidal estuary)	205 acres
Rochester, Maine, U.S.A.	Silver Lake (waterfowl area)	20 acres
Smoke Rise, N.J., U.S.A.	Planned suburban development with retained wildlife habitat	3,500 acres
Washington, D.C., U.S.A.	Rock Creek Park (wooded river valley)	1,754 acres

EXAMPLES OF "PLANNED NATURE"

Experience in the planning of natural habitat for non-game species in urban areas is meager. An assortment of cases is presented here, but in view of the variability of the available information only an outline review is possible. The fundamental principles and working policies employed in each project have been extracted together with any additional information which might be useful to urban wildlife planners. Where possible, examples selected illustrate the different opportunities for planned wildlife areas mentioned earlier.

Preservation of a Remnant Habitat

In the city of Regina, Saskatchewan, a representative natural prairie marsh has been secured and now forms a sanctuary, the Wascana Waterfowl Park, for waterfowl and other wetland species. A marsh and lake area of some 360 acres is managed to provide suitable grounds for breeding, feeding, and molting waterfowl. The density of birds is higher than would be expected in the wild condition and, due to the lack of harassment, they are much more tolerant of human presence, which adds significantly to their value as an esthetic resource.

The location of the lake near a small power station has meant that open water is present the year round, making it an attractive location for wildlife (Hochbaum 1965). The objectives in creating this park were to:

1. Establish and increase bird and other wildlife populations.
2. Retain and develop the marsh in as natural condition as possible, utilizing the associated native flora.
3. Encourage public use of the area to meet recreational, educational and research needs.

To achieve these objectives the following management policies were adopted:

Wascana Waterfowl Park in the city of Regina, Saskatchewan, is a natural marsh that provides a large breeding ground for wildlife within an urban area. (*Photo courtesy of the Saskatchewan Museum of Natural History, Regina.*)

1. Establish zones for various uses and activities.
2. Restrict research to that which in no way damages or disturbs the natural conditions.
3. Introduce a "natives only" policy in all planting.
4. Restore certain areas to increase their stability by controlling erosion through bank grading, and provision of loafing sites.
5. Make ecological surveys of all existing plant communities.
6. Use strategies to minimize disturbances and damage caused by human visiting. Such devices as hedges, water barriers, boardwalks, observation posts and hardtop paths can be effective.
7. Minimize the use of chemicals within the reserve (although it appears inevitable that some chemicals will be carried in from outside).

Planned New City Employing the Cluster Concept

The new city of Columbia has been developed in the State of Maryland near the national capital. In 1969 the Bureau of Sport Fisheries and Wildlife of the United States Department of the Interior was asked to investigate the possibilities of planning to retain wildlife habitat within this development. The total area covers some 15,000 acres and of this 3,200 acres are to be secured as green space (Larimer 1969; Wilmers 1971). Principles of the plan are as follows:

1. An attempt is made to devise a land use plan which avoids the necessity of imposing numerous regulations on people using the area.
2. In view of the inherent richness of bird life in the district, it is recommended that mature and overmature forests, under-mowed grassland, fallow fields, bushy hedgerows, and cultivated croplands be perpetuated. Appropriate management practices will be carried out.
3. Land use plan allows areas for low-intensity activities (such as golf course, university, industrial park, and playing fields) to fringe on natural areas.
4. Attention will be directed to solving an anticipated problem, namely, the detrimental effect of intense use of the area and the impact this will have on fragile natural communities.
5. Protection and preservation of all permanent streams is being instituted, with strict attention being given to retaining riparian vegetation.

Certain strategies are employed to increase ecological diversity:

1. Clearings (one-fourth to one-half acre) in woodland and bushy edge on the borders of woodlands and open fields have been created.
2. At least one water area has been developed to provide fishing and to facilitate the establishment of a marsh.

In Columbia, Maryland, a new, planned city, the banks of this stream are left in a semi-natural state. The heavy growth of vegetation provides food and shelter for wildlife. The land on one side is in a nearly wild state; the other side of the stream is mowed and used as a picnic area. (*Silvan photo.*)

The strategies recommended to minimize damage are:

1. Vehicular access to watercourses is restricted to one point only.
2. Prohibition of all motorized boats.
3. An education center for interpretive purposes, located so as to relieve pressure on the most fragile areas.
4. Nature trails and observation blinds will be established at appropriate points.

The proposals for management are:

1. Use of four-year rotation in mowing the grassland area.
2. Maintenance of small units of agricultural land growing traditional crops.
3. Minimal care of woodland; no clearing of debris and leaf litter.
4. Limited use of chemicals.

A Conservation Plan for an Estate

A new estate to include housing and industry is being built in the Estover area of the City of Plymouth in Devon, England. The site, approximately 1,300 acres, now comprises woodland and agricultural land. A plan which attempts to retain, add to, and perpetuate landscape features important to the ecosystem has been presented to the city by the Devon Trust for Nature Conservation (DTNC) (Smith and Wheeler 1970).

F. G. Wheeler (1969:853) believes that the Estover development "clearly presents the possibility of including the conservation of the varied flora and fauna within the site and an opportunity to plan a unique experiment in co-existence between man and wildlife." Principles of the plan are that:

1. Urban environmental quality can be enhanced through the integration of green space into the development during the planning stage.
2. It will be necessary to incorporate restrictions into the plan which will minimize sources of pollution in the new development in order to preserve ecosystems within and beyond the Estover site.
3. There is a paramount need for the retention and provision of biological corridors to facilitate dispersal of animals throughout the area.

The strategies to maintain wildlife are:

1. Hedges will be retained wherever possible, together with at least a ten-foot strip on each side to support under-mowed grassland.
2. All roads will be constructed with 50-foot margins planted to maximize diversity, using hardwood species with a conifer nurse crop.
3. The existing woodland will be extended into the heart of the development area by new plantings to give a continuous chain of wooded valleys separating the open spaces. It is recommended that hardwoods be planted in a matrix of conifers to provide good wildlife habitat.
4. Management practices such as mowing, cutting, and felling will be left to September when they will create the least disturbance to natural populations. Maximum floral diversity will thereby be encouraged and smaller species will be favored. It is suggested that the traditional practice of hedge laying be continued on a twelve to fifteen year rotation. (Hedge laying is a technique for restricting the height of the vegetation and creating a dense, impenetrable barrier through the use of hand tools to partially cut stems of some of the shrubs and direct their growth in a horizontal direction.)
5. Creation of ponds should be attempted in selected areas.
6. The development of an arboretum and insect garden on the site of a demolished house set in an overgrown garden is recommended.
7. Playing fields will form part of the biological corridor network, and care will be exercised to ensure that native vegetation is retained in such locations.

A New University Campus Developed to Enhance the Natural Landscape

The landscape architect Neil Porterfield (1969) recently produced a plan for the new Parkside Campus at the University of Wisconsin. His approach to the project included the following steps:

1. A detailed inventory together with site analyses of all natural and man-made features was conducted prior to the drawing up of the development plan.
2. A study of the ecological factors influencing the existing landscape was made.
3. Natural features of high esthetic value (river valley, flood plain, and woodland) were designated for preferential retention.
4. All suggestions for planting and landscaping were based upon the principle of producing an area requiring only minimum maintenance, as exemplified by the following statement:

It is our premise that a landscape planned in accordance with and maintained by practices simulating nature's own processes will reduce maintenance. In contrast, horticultural treatment of institutional land holdings requires costly weekly mowings, frequent fertilizing, and pruning of introduced grasses and exotic species.

Roadside Verges as Wildlife Habitat

Wildlife inhabiting green areas that border roads in Britain has been discussed by Way (1970). Since the area of such land is in excess of 10,000 acres, studies are being conducted to examine the existing plant communities in roadside habitats and to assist in the formulation of management plans. It is essential that these areas be managed to develop stability since they are likely to become an increasingly important reservoir of wildlife. Toxic concentrations of automobile pollutants that can build up in such locations must be considered when attempting enhancement of roadside wildlife communities.

In the County of Kent, a co-operative arrangement between the County Council and the Department of Horticulture of London University has resulted in an experimental planting of new roadside cuttings to simulate natural cover. This project is being conducted in an area of high landscape value and thus reestablishment of typical vegetation was considered essential. The method required the cutting of turf (regular-shaped pieces of soil and vegetation 12 x 12 x 6 inches) from adjacent habitat areas and planting them in two lines two and a half feet apart near the top of the bank. Turfs of ling (heather) (*Calluna vulgaris*), bilberry (*Vaccinium myrtilis*), and bracken (fern) (*Pteridium aquilinum*) were used and eventually a random block planting of the whole slope was accomplished. A review of the project showed that a 70 per cent establishment was achieved and the technique of planting two lines at the top of the bank successfully stabilized topsoil and minimized erosion. Rapid colonization of the bank by herbs and shurbs occurred between the turfs presumably as a result of the protection and the adequate supply of seeds and spores in the immediate area.

Restoration of Derelict Common Land, England

Mitcham Common is a 400-acre open space in London which has been severely abused in recent decades through its use as a refuse tip. The London Borough of Merton has plans to reinstate natural vegetation resembling that which originally existed (H. P. Boddington, Director of Parks and Cemeteries, London Borough of Merton, 1971, personal communication). The area chosen for the pilot experiment is some 20 acres in extent. By suitable contouring, addition of topsoil, seeding, planting, and careful management, it is hoped that the picturesque quality of a sandy heath can be reproduced. The procedures employed are:

1. An 18-month survey to assess the various attributes of the entire area and to choose a pilot site.
2. Application of an acid topsoil after re-contouring of the refuse tip and its clay covering.
3. Planting of saplings of native species such as oak and birch, and of an introduced pine (*Pinus sylvestris*), and encourage-

ment of typical shrubs such as blackthorn (*Prunus spinosa*)
and common buckthorn (*Rhamnus cathartica*).
4. Encouragement of natural colonization by willow and elder.
5. Within the larger area of the sandy heath, encouragement of
two principal plant communities—grassland and woodland.

It is uncertain whether it will be possible to simulate the
indigenous vegetation cover, since the drainage conditions
are not typical of those found in undisturbed sandy heath.
To maintain the newly established cover, all grassland will
be mowed three times each year to simulate grazing.

Summary

This assortment of projects illustrates how planned natural habitat can be incorporated in a variety of land-use
plans. The expertise gained in these particular localities
should be drawn upon to encourage similar endeavors
elsewhere. In the majority of cases, except the Wascana
Waterfowl Park, the plans are still new or only partially
implemented and it is too soon to assess their quality.
Each one deserves a more detailed appraisal to evaluate the
objectives, the techniques employed, and the results obtained.

These projects do serve to emphasize the need for important natural sites to be identified prior to any development so that site integrity may be safeguarded by wise
planning of adjacent land use. Only minor changes in
present management techniques can considerably enrich
habitat diversity. It is strongly recommended that the
practice of habitat enhancement be incorporated in all
maintenance of public green space.

CHAPTER SIX

Managment of Urban Wildlife Habitat

Management of Urban Wildlife Habitat

Total land-use planning such as outlined in Chapter Five, which includes provision of wildlife habitat, is unlikely to prove rewarding unless an equally great effort is devoted to refining land management techniques (Duffey 1967). To manage wildlife populations it is necessary to manage habitat. Natural systems are dynamic; they change through time toward a higher degree of complexity unless counteracting forces intercede to prevent the successional modification of the environment. Management may therefore be viewed as a means whereby regression or progression of an ecosystem is altered to some predetermined stage. It is possible to manipulate physical factors such as drainage and water levels or to modify the complexity of the system by adding or removing energy and materials; management is in fact a form of applied research (Nicholson 1956).

MANAGEMENT OBJECTIVES

Management may be needed to protect and maintain an existing habitat or it may be used to shape the development of habitat in an area. There is a great difference between these two objectives and this will determine the policies employed to achieve the desired end.

In all management activities one principle which should be adhered to is the maintenance of the highest degree of diversity which is compatible with the system to be perpetuated. In the case of an early successional stage where diversity will be minimal, the degree of management activity required will be most intense (Streeter 1968). Diversity is associated with stability, therefore it should form the principal focus for management of habitat in urban areas.

POLICIES

Diversity can be enhanced by several means. Within any single community it can be achieved by encouraging the establishment of as large a variety of native species as possible. It can be further promoted through the *in situ* retention of dead and dying organic material (Manns 1967). The structure and nature of the vegetation should similarly be diversified since they are major determinants of animal density (Beecher, 1942; Colquhoun and Morley 1943; Dowdy 1947; Preston and Norris 1947; MacArthur *et al.* 1962; Beven 1964; MacArthur 1964; Hooper and Crawford 1969; Recher 1969).

In pursuit of variety, an alternative to the manipulation of a single community is the creation of adjacent ecosystems of varying age such that all phases of succession are represented in a region (Tischler 1956; Odum 1969). This approach would necessitate the rejuvenation of selected portions of the mature systems to ensure a continuity of heterogeneous units. This technique would provide continuity through time in any one area, but at any specific location there would be a gradual development toward larger biomass, lower productivity, and greater stability. Given adequate linkage between habitats, mobility and capacity for dispersal of animals and plants would ensure that these intentionally disturbed habitats were colonized.

In densely populated countries the landscape is either seminatural or cultivated and the wildlife populations are adapted to these conditions. In order to retain these communities there is a need to preserve the cultivation practices that have created them. Examples of such practices are the laying of hedges, coppicing of woodland, grazing of grassland, and cutting of reeds in marshy areas. As many varied land uses as possible should be encouraged so as to present a mosaic of habitats and to encourage the diversity of species that ordinarily occurs where different types of habitats meet.

TECHNIQUES OF DIVERSIFICATION

Topographic Modification

Landscape can be changed by remodeling the ground surface to create hills and depressions which will add greatly to the range of potential habitats (Jellicoe 1966; Weddle 1967; Colvin 1970; Fairbrother 1970). Techniques for excavating ponds and lakes are well known but some are less appropriate than others. The successful method known as "pothole blasting," for example, is inapplicable for built-up areas (Provost 1948; Uhler 1956; Chabreck 1968; Hoffman 1970), while earth-moving equipment can create shallow lakes and canals with little difficulty. It is quite common to find in the vicinity of populated areas numerous worked-out gravel and clay pits or quarries; and these with judicious planting of some shrubs and food plants, and creation of gravel bars, will support waterfowl populations (Harrison 1964; Olney 1964; Glue 1970). In larger ponds and reservoirs additional niches can be established by supplying rafts and logs to create artificial islands and also through forming bars and peninsulas (Sugden and Benson 1970).

The use of small earth dams in areas of high water table can create suitable conditions for bog and marsh vegetation, both of which are biologically productive (Bradley and Cook 1951; Brumstead and Hewitt 1952; Boyer and Devitt 1961; Shomon *et al.* 1966; Niering 1968; Bye 1969; Smith 1970).

Increasing Species Diversity and Density

Variety may be encouraged by seeding and planting vegetation. In recent years techniques have also been devised for transplanting mature trees, thus allowing an apparently mature system to be created in a short time. The subject of city trees and their culture has been considered by several authors (Knechtel 1909; Min. H. L. G. 1958; Salter 1960; Haddock 1961; Marling 1963; Civic Trust 1966, 1967; Zion 1968; Hebblethwaite 1969).

Seeds or saplings for urban planting should be drawn where possible from local populations in order to maximize their chances of establishment. All plantings should be mixed, with different species of both deciduous and evergreen form (van Dersal 1941; Kubichek 1941; van Emden 1965; USDA 1969). The presence of a variety of trees ensures that cover will be available earlier than if a single late-developing species were planted. This is very important since it influences songbird density and breeding success. The increase in vegetation diversity and stratification can bring about an automatic and complementary enhancement of animal variety, especially if indigenous plants are used. Where it is thought desirable to encourage certain animals in the urban environment, an attempt should be made to incorporate their preferred food plants

A nearly barren area in a flooded gravel pit in west Kent, England, has been made into a highly productive wildlife area. Above: A shallow inlet is created. Top right: The next year food plants are well established. Bottom right: In four years a rich and varied plant life has developed. All three pictures are of the same site in a 70 acre gravel pit. In 1960–61 (first photo) only 5 species of birds and a few dozen individuals were reported here. By 1964–5 (last photo) 10 species were seen and thousands of individual birds were counted from August through April. (*Photos by Pamela Harrison, A.R.P.S.*)

into any vegetation plan. There is some literature on this subject and much more is likely to accumulate with the growing number of ecological studies in different communities (Collinge 1936; Jones 1940; Martin *et al.* 1951; Hartley 1954, 1966; Owen 1956; Olney 1958, 1967, 1968; Davison and Hamor 1960; Longnecker and Ellarson 1960; Davison and Grizell 1961; Parr 1963; Dasmann 1964; Davison 1964; Gill 1965; Kendeigh and West 1965; Robel and Harper 1965; Davis 1967b; Pollard 1967; Krull 1970).

It is possible to increase animal density by directly providing food materials, but this can lead to severe oscillations in numbers if any discontinuity occurs in the food supply. Attempts to increase animal numbers should therefore be accomplished by establishing food-bearing plants in a suitable spatial arrangement (Nicholson 1956).

On a long-term basis it will be necessary periodically to plant trees to ensure retention of uneven-aged woodland with a maximum number of niches. In order to achieve this, small stands should be felled with the creation of areas of freshly turned soil to aid the establishment of locally available seedlings. It is debatable whether the dominant trees in an established forest are always capable of replenishing their numbers; the state of the soil surface may be a critical factor. Mellanby (1968) reports successful germination of oak seedlings in recently cultivated ground adjacent to mature woodland but within the woodland no sign of regeneration was detected. He found no satisfactory explanation for these observations. Intolerance of deep shade is a factor that limits regeneration within some mature woodlands, and this may have had some influence.

There is considerable disparity of opinion among conservationists concerning the advisability of introducing exotic species (Elton 1958; Yapp 1966; de Vos and Petrides 1967). Disruption of the ecosystem may be inevitable when foreign organisms are allowed to enter a natural community. Howard (1965, 1967) states that our knowledge of the factors which regulate productivity and stability of ecosystems is still insufficient to warrant deliberate introduction of exotic species for whatever purpose. This question has been discussed recently by Lewin (1971) in connection with the Hawaiian Islands which appear to have been subjected to more introductions per unit area than any other part of the world. He notes that exotics pose a threat to native species particularly where they happen to be closely related, since hybridization can occur.

One of the reasons that mature systems possess stability is the intricate meshing of the interactions between species which serves to contain the size of any population. These controls have evolved over millennia and it is therefore not surprising to learn of the population explosions resulting from certain introductions (Elton 1958).

In the United States it has been common practice among wildlife managers to import game species, but the management effort required to maintain many of these animals has been very great (Dasmann 1964; Gottschalk 1967). Artificial breeding and subsequent release is another game management technique which seems out of place in managing natural populations in urban areas. If it is possible to establish reserve areas of viable size within the green space network, such practices should be superfluous. It is unlikely in any case that artificial rearing can increase animal populations on a permanent basis, since the primary limiting factor is suitable habitat (Boyd 1957).

In North America, as previously mentioned, a considerable volume of literature has been assembled pertaining to game species but little attention has been directed to nonconsumptive wildlife management (Giles 1969; Hay 1969; Hooper and Crawford 1969; USDA 1969).

As a general method, increasing the number of nesting sites has proved to be successful in enlarging animal (especially bird) populations (Brown and Bellrose 1943; Stuewer 1943; Terres 1953; Craighead and Stockstad 1961; Staude 1966; Cohen 1967; Busse and Olech 1968; Davey 1969; Kalmbach and McAtee 1969; Sawyer 1969; Bishop and Barratt 1970; Will and Crawford 1970). In city environments the deliberate provision of brickwork and masonry with various sizes of apertures can give shelter and breeding space to songbirds, owls, and bats. In Schlangen, West Germany, 200 new houses were constructed that included built-in nesting boxes, and these have attracted a variety of birds which use them successfully. Children's adventure playgrounds offer excellent potential as nesting areas and this could be realized by the modification of some of the equipment provided in these areas. Strawinski (1963) believes that the most important factor limiting bird populations in the city environment is the paucity of nesting sites. Fortunately, nesting sites can be provided where desired at a minimum expense and with a high degree of success.

CROPPING AND CONTROL TECHNIQUES

Numerous methods are known for the control of vegetation; these range from manual through mechanical to chemical and biological. Each operates in a different manner and, depending on the objectives of the management plan, some are more appropriate than others.

Many traditional cultivation practices have produced the landscape we view today. Grazing has been effective in maintaining chalk grassland in southern England (Floyd 1965; Wells 1965). Pollarding of willow trees has increased the number of hollow and dying individuals, thus creating many niches (Glue 1970). Coppicing is a forestry practice, whereby the trees are cut down to within 15 inches of the ground every 15 years, which tends to maximize the biological richness of woodland (Rackham 1967; Carlisle and Brown 1967; Welch 1968). Burning of moorland has been the major tool for the managers of grouse (*Lagopus lagopus*) in Britain, and fire is also useful in preserving early successional phases (Picozzi 1968; Lindsey County Council 1970). However, burning of grassland to clear scrub is considered detrimental to the maintenance of a diversity of flowering plants in the burned area (Lindsey County Council 1970).

In towns, in any case, it is increasingly unacceptable to use burning as a management technique, owing to the resultant air pollution and potential hazard to property. To maintain wildlife populations in commercially forested areas it is necessary to retain some mature vegetation within the felling zone, and the brushwood resulting from forestry operations can provide excellent cover for wildlife if left *in situ* (Williamson 1967), just as fallen trees can in urban parklands.

In connection with urban wetland it is unfortunate that nutrient inputs are usually high and thus some removal of materials will be needed to prevent succession toward a non-aquatic system. Repeated clearing of emergent vegetation and occasional dredging may be necessary and this should be done manually to avoid more than minimum damage (Clawson 1969; Lindsey County Council 1970; USDI n.d.). A review of the problems attendant upon removal of riparian vegetation is provided by Campbell (1970).

With the development of chemicals that act as growth inhibitors or accelerators a new tool was made available to conservationists and game managers. There has been a general acceptance that such chemicals are non-persistent and relatively harmless to wildlife, provided they are used in accordance with recommendations (Allen 1953; Way 1969). In recent studies of game birds in France, however, there has been some evidence for mutagenic effects upon embryos, and this raises the question of the long-term effects of such chemicals on ground-nesting birds (New Sci. 1970, 49:593).

Use of herbicides has been documented in relation to many non-agricultural land uses (Ibberson 1951; Egler 1952, 1954; Pound and Egler 1953; Bramble and Burns 1955; Goodwin and Niering 1959; Hopkins 1961; Niering and Goodwin 1963; Ward 1964; Kenfield 1966; Newton 1967). In view of the current uncertainty about the safety of herbicides and the known persistence of many available pesticides, their application should be regarded as a last resort in managing urban wildlife habitat. In parks and gardens it is quite unnecessary to employ chemical controls, since there is no requirement to grow perfect crops or flowers to fit a Madison Avenue image (Fowles 1970). Unlike agricultural land, many urban gardens comprise a mosaic of small-scale habitats and, because of the large number of niches, normal biological controls work well (Mellanby 1967).

Control of Animal Populations

The problems involved in controlling sparrow, pigeon, and starling populations in urban areas have been reviewed by Thearle (1968). As he points out, the most difficult aspect of the problem is the attitude of the public to any such culling. Neither cage trapping nor shooting was deemed permissible, but use of narcotic baits held promise of being adequate and inexpensive if used against discrete populations (Ridpath *et al.* 1961; Murton 1965). By conducting the treatment at dawn the probability of public reaction was minimized. The greatest success has been with sparrow populations; this was attributed to their attachment to specific colonies and the fact that only juvenile birds disperse to new breeding sites (Summers-Smith 1963).

Superabundance of any one species in an ecosystem can usually be attributed to ecological mismanagement, and this can only be remedied by modifying the environmental conditions so that the undesirable population is placed at a disadvantage (Lawrence 1967). A reduction in carrying capacity is thus required to bring into play mechanisms such as competition and predation which will increase mortality and in some instances may reduce natality (Davis 1953). It is debatable whether our knowledge of habitat requirements is sufficiently advanced to enable such an approach but it is worth exploring further. For example, urban pigeon populations can be held down by limiting the number of ledges suitable for the birds' nesting. Traditional control methods attached great value to the concept of extermination but, as has been pointed out by Cain (1966), if successful it merely creates a vacant niche which may then be colonized by yet another undesirable species.

PROBLEMS ASSOCIATED WITH CERTAIN MANAGEMENT TECHNIQUES

Not all established methods will be transferable to the urban setting. Grazing for instance can be a most useful tool for the maintenance of short grass habitat, but in urban areas there is likely to be a problem with the number of dogs which could harass sheep (Reynolds and Sankey 1967). This problem has been overcome in Edmonton, Canada, by proper fencing. Here the University of Alberta Farm safely maintains large numbers of sheep within the city. Cattle might also be used as a management tool, but hobbling or temporary fencing would be required so that only chosen areas would be grazed. If allowed free access, these animals cause damage to fragile vegetation and may reverse succession to an undesirable degree (Dambach 1944).

Seasonal Use of Techniques

Where the primary objective of management is the maintenance of wildlife habitat, it is essential that care be exercised in the timing of all manipulation activity. An understanding of the ecology of each species is necessary in order that disturbance can be minimized. In maintaining grassland, for example, mowing should not be allowed to interfere with ground nesting birds. In addition, there should be monthly variation in cutting to avoid destroying the same plant species each year. Other activities such as thinning, felling, and pruning should also be timed to avoid the breeding season of birds and other animals if possible. Considerable difficulty may be experienced in deciding upon the "best" time, since many shrubs and trees furnish food and shelter all year round. In general, most management activity should be concentrated in the late autumn and winter months.

PROBLEMS FROM HUMAN USE OF
NATURAL HABITAT AREAS

Damage

Trampling of vegetation is inevitable in any regularly used area and, depending on the substrate, varying degrees of damage occur. On chalk grassland, for example, three levels are observed: light trampling, which has little effect upon the plants; medium trampling, which abrades the plants and results in a close-cropped turf and a reduction in species diversity; and the third level, where plants are completely killed, often resulting in soil erosion. Horse riding has even more detrimental effects upon vegetation cover, particularly where soil moisture is high, as in woodlands and along stream valleys. Cars and "recreational vehicles" often intrude upon the vegetation of natural areas with disastrous results (Perring 1967a; Reynolds and Sankey 1967; Lindsey County Council 1970).

Collection of wildflowers and removal of whole plants is a great problem in the vicinity of populated centers. In southern England many plants have been extirpated from the countryside surrounding major settlements. The common primrose (*Primula vulgaris*), for example, is now rare within a 30-mile radius of London (Lindsey County Council 1970).

Accidental fires are frequent in heavily used recreational areas and these can have disastrous effects upon both flora and fauna. In England, bracken (*Pteridium aquilinum*) has acquired dominance on many acres of heathland due to its ability to withstand fires through its rhizomatous habit (Watt 1955).

Disturbance

The ability of many animal species to tolerate human disturbance is not well developed, and this is a limiting factor of considerable importance in urban areas. The type of activity which may be classified as disturbance includes deliberate nest destruction, collection of eggs and young animals, and harassment and injury. It has been suggested, for example, that kestrels, which in England have taken to living in suburban areas, may be very vulnerable to nest predation in view of their popularity among falconers (Lond. Bird Rep. 1968, 32:85). Often destruction is unintentional and stems from ignorance of the living system of which vegetation forms the foundation. Such occurrences as the breaking down of lakeside vegetation by fishermen in order to secure a suitable fishing spot and the beating down of undergrowth by inept golfers are typical problems.

Shooting is illegal in most urban areas but air guns are popular with children and may be a serious problem in localized districts where isolated wildlife populations could be endangered. During the breeding season freedom from repeated disturbance is essential if young are to be successfully reared. It is for this reason that sanctuary areas are indispensable in the urban environment. The mere presence of people can drive some animals away from one small but essential part of their territory, thus forcing them to abandon the area entirely. In southern England bats have been disturbed due to people visiting caves which have served as roosts for centuries (Lond. Nat. 1968, 47:43).

The distribution of litter and refuse can have marked effects upon plant growth and animal mortality. Materials which are not decomposable distort plant growth or kill individual specimens; discarded containers can act as lethal traps for small animals and invertebrates (Harper and Morris 1965). Another result of the influx of large numbers of people to a natural area is the increased probability of introduction and removal of spores, fruits, and larvae on clothing and footwear; these may be transferred between widely separated locations by motor vehicles. A report of studies conducted on a heavily used piece of downland in Sussex shows that noticeable inputs of nutrients to soil occur as a result of human use and this fertilizing effect alters both species composition and luxuriance of the vegetation (Lindsey County Council 1970).

Urban natural areas are unduly influenced by their location; there is an obvious need for techniques to relieve some of the pressure of use. If stable natural communities are desired, it will be necessary to manage people in addition to the other components of the ecosystem.

DISTRIBUTION OF PEOPLE IN
RECREATIONAL AREAS

In order that the management of people can be most effective it will be necessary to improve our understanding of the way people use open space. If detrimental effects upon wildlife are to be minimized, the conflict zones between human and wildlife activity must be defined and the habitat manipulated so as to minimize the confrontation. For example, Forman (1968) points out that the places most favored by recreationists are the habitat boundaries such as waterside, forest edge, and hillsides—which are also those most useful to wildlife.

People tend to stay relatively close to the main points of access to any natural area. Highways are the principal focus for recreationists; in metropolitan parks most people stay within one hundred feet of their parked cars (Satterthwaite and Marcou 1969). It is not easy to predict whether such limited penetration will similarly apply to natural areas situated in residential districts. The presence of a high percentage of children playing in a "known" rather than a "strange" environment is likely to result in a more extensive use of the area and in consequence the disturbance of wildlife will be greater.

"PEOPLE MANAGEMENT"

Controlling Access

There is considerable evidence for the benefits of establishing certain parts of any area as a sanctuary where wildlife can remain relatively undisturbed (Lond. Nat. 1965, 44:158; Grant 1967; Lond. Bird Rep. 1967, 31:65; 1968, 32:68). In London's Kensington Gardens, for instance, the largest increase in bird numbers and species type has been reported from the Long Water Sanctuary which occupies a mere two acres within the total area of 275 acres (Lond. Bird Rep. 1970, 34:55).

The manner of controlling access may be varied according to the local situation, but careful site planning can provide a built-in zoning by making some areas more inaccessible than others. Water can exercise a strong inhibition to entry; thus islands automatically become sanctuaries, provided boating is not allowed. In woodland, key areas can be fenced off, leaving paths between to allow people visual but not physical access. To protect fallow deer (*Dama dama*) in Epping Forest on the outskirts of London, enclosures have been established with "deer leaps" to allow the adult deer freedom to come and go as they please while at the same time restricting the exit of young animals. Fawns born within the enclosure are unable to leave this area until they are at least one year old, and this has considerably reduced the death toll on adjacent highways (Lond. Nat. 1963; 42:101; Chapman 1969).

An inexpensive barrier to human access often more effective than fencing is provided by thorny shrubs. Such species as bramble (*Rubus* spp.), blackthorn (*Prunus spinosa*), gorse (*Ulex* spp.), and holly (*Ilex* spp.) will inhibit human entry while providing excellent cover.

Prevention of the entry of cars can be achieved through use of various devices. Adequate car parking space in a carefully chosen location should always be the prime objective. In addition, the blocking of footpaths and trails by earthen ramps, fences and gates, or water-filled canals, can control access of cars; but frequent patrols are usually necessary to prevent entry of vehicles such as "off-the-road" motorcycles into natural areas.

Increasingly it has been realized that restrictive controls alone can never provide adequate protection in the face of intensive human use, and the popularity of nature trails has provided one positive means of controlling the areas of

---

Natural islands (or man-made ones) can be planted with native trees, shrubs, and herbaceous plants that provide food and cover. In lakes and rivers, even in the middle of cities, such islands then become valuable sanctuaries for wildlife. These two photos show a few sandy islands in England in 1965 before they were planted, and in 1968 when plants had grown. Six different species of birds were nesting on these islands when the second picture was taken. (*Photos by Pamela Harrison, A.R.P.S.*)

use. Trails may be guided or self-guided, but either way they reduce the tendency to wander randomly and thus substantially reduce the disturbance to wildlife. Careful planning of the trail layout is essential and should have the objective of directing people away from the areas of greatest importance to breeding animals.

Education

In the final analysis, the most rewarding form of people management lies in an educational program designed to create positive attitudes toward natural resource conservation. The primary aim of any program should be the encouragement of an awareness of the human role in the biosphere. Particular stress should be laid upon the responsibility of each individual to press for the establishment of a stable, contained system which optimizes energy and material usage, and satisfies as many levels of human needs as possible.

Education should be regarded as an indispensable tool which can contribute to the protection of urban wildlife habitats by ensuring that they are not used beyond their ecological carrying capacity.

Making wildlife more accessible to the public while at the same time providing adequate protection from disturbance is likely to prove the most challenging problem for the urban resource manager. Accessibility of wild animals can be greatly enhanced by providing nature trails with strategically located observation blinds equipped with suitable optical aids (Wildfowl Trust 1956). In the city there are many opportunities for the installation of observation posts in such locations as parks, golf courses, gravel pits, and reservoirs.

Involvement of local people in the management activities of natural areas is one way to encourage an enlightened interest, and is particularly applicable where children are concerned. Rawley and Peucker (1970) record that children use parks to a much greater extent than do other age groups. When reclamation of coal-mined areas was first attempted in Britain, youthful vandalism was the greatest deterrent to success of the planting programs. Realizing this, the administrators undertook an information and education campaign in the schools and involved children in the replanting scheme, which significantly improved the establishment of cover. Shomon (1971) mentions cases in the United States where vandalism has been controlled by involving young people in nature center programs.

To integrate the educational programs, interpretive centers should be established in the principal green spaces to provide information and expert help in the enjoyment, understanding, and use of these areas.

MANAGEMENT OF URBAN GREEN SPACES

City administrators rarely show adequate understanding of ecological principles pertaining to habitat management. Indications of an enlightened attitude to planned development of natural landscapes are also commonly lacking. Typical abuse of open space is its use as a repository for all forms of litter and waste; many biologically rich locations have drowned under a welter of garbage. In London there are numerous examples of ponds on common land being "reclaimed" by filling with rubbish (Lond. Nat. 1962, 41:25; 1967, 46:24; Mitcham Common Survey 1967).

Abuse of vegetation also takes the form of destruction of old timber, pruning without appreciation for the growth form of the tree, and the placing of notices and brackets on and around trees which damage the bark and allow pathogens to enter the tissues. Timing of management operations rarely takes into consideration wildlife requirements and much damage results (Lond. Nat. 1961, 40:22; 1962, 41:66; Lond. Bird Rep. 1963, 26:79-80).

Too often the principal concerns of park managers are pest control, reduction in property damage, and provision of flamboyant and temporary displays of exotic species, all of which involve high energy input and large expenditures of funds and man-hours. The result is a costly and highly unstable system.

Roadside borders suffer from heavy-handed management which neither enhances their appearance nor their usefulness as wildlife habitats. Repeated blanket spraying eliminates colorful flowers and produces a monoculture turf (Egler 1954; Perring 1967b; Way 1967). A study by Oetting and Cassel (1971) illustrates the benefits of appropriate roadside management.

Proposed Guidelines for Urban Open Space Managers

Objectives: To encourage multiple areas that support viable populations of wild species within the urban matrix.

Policies and Guidelines:

1. Planned development of any kind within the city should provide for wild areas.
2. Retention and creation of biological corridors should be given priority in all development plans to maintain the viability of all open spaces as wildlife habitat.
3. The importance of linear habitat such as hedges, verges, utility rights-of-way, stream banks, and walls should be recognized, and these should be retained. Efforts should be directed to the management of such habitat to maximize its biological potential and enhance its function as a corridor.
4. Attention should be paid to increasing diversity wherever feasible by such means as:
   a. interspersion of different land uses

    b. planting of mixed species to create multistratified vegetation cover

    c. retention of decaying vegetation and animal matter *in situ.*

5. In all management plans rejuvenation of some habitats on a periodic basis should be accomplished by small-scale disturbance over limited areas, such as the clearance of openings in woodland, thinning of shrub, removal of emergent vegetation around small ponds, and the cutting of grassland every three years.

6. Seasonal operations should be organized to cause minimum disturbance to wildlife activities; i.e., concentration of activity should be in late autumn and early spring. Where possible manual rather than mechanical techniques should be used.

7. The use of chemicals should be abandoned in favor of other forms of vegetation manipulation. Fenced enclosures should be established to encourage reestablishment of plants rather than relying upon the use of fertilizers.

8. Sanctuary zones for wildlife should be established which can then function as reservoirs for the remainder of the green space network.

9. The high biological value and protective function of wetlands such as bogs and marshes should be recognized. These areas should be protected from garbage dumping and from drainage schemes or other forms of detrimental activity.

10. Acceptable modes of people management should be developed to create a general awareness of the ecosystem and its functioning. Damage should be minimized by restricting use to a level below the ecological carrying capacity.

11. The rotation technique should be adopted as a major protective tool in all areas subjected to intensive human use. For example, different portions of a parkland should be periodically fenced to exclude people and allow recovery of vegetation. Alternative walking trails should be provided to maintain each one within its ecological carrying capacity by periodic enclosure.

CHAPTER SEVEN

Conclusions

When this study was first conceived, the authors presumed that little attention had been paid to the subject of urban wildlife ecology; and this assumption has been confirmed. It is true that specific topics within the general subject area have been investigated, for example lichens in cities, suburban songbird ecology, and control of urban pest species; but no evidence has emerged of efforts to synthesize the available material. In North America until very recently there has been undue emphasis upon consumptive wildlife management which goes some way toward explaining the basic problem encountered during our study—namely, a dearth of published facts. Much of the more valuable information still resides with local conservationists, many of whom are nonprofessionals and therefore not exposed to pressure for publication of their findings. This leads to a most unfortunate situation where opinion is widely available (USDI 1969), but solid information is largely unobtainable.

In the light of the information acquired, the questions posed in the Introduction may now be reviewed. The first of these considered the existence of urban areas which contain viable natural communities. The description of

London and Los Angeles and the analysis of factors contributing to their role as wildlife areas provide evidence that natural communities can be maintained in such places. Further investigations in other areas could be expected to reveal other locations where similar natural elements of the cityscape persist.

The second question on the subject of values is essentially unanswered since this aspect of urban ecology is less amenable to investigation. In many of the sources consulted various psychological values are attributed to the presence of nature in the city but few data or references are cited. This appears to be an interesting facet which should be explored by both sociologists and psychologists whose information could be useful to planners and perhaps to the medical profession. Involved in this question of esthetic values are the additional economic implications; for example, how much do greenery and songbirds influence people's location decisions? Potential areas of research are suggested by the following questions:

1. Is there any evidence for positive effects upon health and mental well-being as a result of having regular contact with natural phenomena, living organisms, and natural processes (i.e. with "nature")?

2. Does the presence of wildlife in close proximity to houses create an awareness in people which is then reflected in the value they place upon such esthetic resources? Does it in any way make them more aware of their role in the ecosystem?

3. Is it possible that awareness of nature in the vicinity of the home will alert people to more distant conservation issues; for example, the retention of wilderness areas?

4. What factors influence a person's perception of wildlife?

5. What practical value with respect to pollution control and reduction can vegetation play in the built environment? What are the medical implications of this?

6. How much of the monetary value of property stems from

the form of the surroundings? Does the presence of a mature and diverse residential landscape raise the desirability of a neighborhood in any quantifiable way?

Turning to the third of our original questions which dealt with the special characteristics displayed by urban organisms, it is apparent that some ·groups of plants and animals are more prominent than others in urban areas. Omnivores are prevalent, as are nocturnal mammals and those capable of flight; in fact, any organism that shows adaptive behavior can potentially thrive in the urban environment. These observations lead to the conclusion that creatures of the city will be those normally associated with early plant successional stages where a premium is placed upon features such as adaptability, high productivity, and rapid turnover—unless, of course, some provision is made for biological diversity in the environment. The possibility of highly specialized animals establishing themselves is exceedingly low because of the inherent perturbations of the urban system.

Leading on from this is the consideration of the limiting factors that act on urban wildlife populations. Man's activities which lead to habitat destruction constitute a great and powerful force acting on both urban and rural populations. Furthermore, our increasing contamination of the environment bids fair to extirpate many species which were formerly considered abundant, for instance the common frog and the partridge (*Perdix perdix*) from both the city and the countryside of Britain.

With respect to such limiting factors there is but one answer, and that lies with the practitioners of habitat management. A fundamental problem exists, however, in the wide chasm that often lies between those who manage urban natural areas and those who possess the information needed for such management. Ecologists must therefore recognize the necessity to turn attention to the numerous and challenging problems of the built environment so that habitat management may move from its present state of speculative exercise to one where controlled experimentation becomes the norm.

Not only ecologists but those in the wildlife management field should consider which facets of the present body of information may be applied in the urban setting. There is a need to assemble and present this knowledge to those who will be confronted with the task of managing the predicted enlargement of natural habitat in tomorrow's cities.

Additional applied ecological research is needed, including:

1. Controlled experiments in habitat creation.
2. Investigations of the minimum size of different habitats in addition to the problem of maintaining the area in the desired successional phase.
3. Investigations of common urban species, to better understand their population dynamics and thus provide the necessary information by which their numbers could be controlled in the event that this becomes necessary. Other aspects of urban wildlife characteristics requiring attention are migration, competition, and adaptation.
4. Investigations of the carrying capacity of different types of habitat in the city.
5. Derivation of techniques whereby this capacity might be raised.
6. A search for techniques to maintain human use within the ecological carrying capacity.
7. Consideration should be given to the disparity between the perception of the natural landscape by the layman and the professional ecologist. This is important since it is the layman who essentially controls the decisions which protect or desecrate natural areas.
8. Derivation of an easily assessed "habitat index" for use by nonspecialists in determining the value of potential habitat.

For planners and developers the importance of retaining natural habitats as integral parts of developments, wherever feasible, should be stressed. Bulldozing a site clean and then following a beautiful landscaping plan to add a green finish to a development is a highly suspect practice. It is eminently more reasonable to retain the natural beauty of a site, and set the buildings within this framework to perpetuate a high quality landscape, than it is to scrape away all vestiges of mature biological form and later try to recreate them in a matter of days. If McHarg (1969) and his followers are correct, ecological analysis as a basis for landscape development and land use planning should be most economical. It behooves planners to study further such schemes as those now in progress at Columbia (in Maryland) and Estover (in Devon) to ascertain whether indeed it is less costly and what kind of problems emerge from the application of this "planned nature" concept.

Some management problems can be expected in the planned communities where communal open space is retained. Unless special forms of administrative agency are established, such communal areas may become sources of strife and potential danger to the human residents in the locale. Planners must therefore take upon themselves the added responsibility of providing after-care advice. Perhaps each development in which significant green space is included should be provided with a detailed five-year management plan to ensure that the original objectives are met.

A problem that challenges those responsible for the management of the city habitats is that of controlling the people who use these areas. Vandalism and ignorance may well represent the greatest threat to any efforts of urban wildlife enhancement. Restrictive measures alone will not provide a cure and positive steps must be taken through the imaginative use of natural barriers and judicious location of wildlife areas with respect to conflicting land uses. Educational facilities and programs designed to create awareness and a desire to comprehend more fully the complexity of nature will provide the other principal tool for protection of habitat.

One key to success in this respect may be the involvement of the public in the management of natural areas. A great deal of talent and knowledge goes to waste at present in North America; this source could be tapped to ensure the well-being of city wildlife populations, as indeed is done in Great Britain where the amateur naturalists are in the forefront of the conservation movement.

Urban ecology demands an interdisciplinary approach if we are to improve the factual basis of decision-making in this challenging facet of resource management.

Glossary

Adaptability — The ability of an organism to respond to changes in its environment so as to continue living and reproducing; no genetic change is involved.

Apothecium — Cup or saucer-shaped fruit body of certain fungi and lichens.

Basidiomycetes — A group of fungi; includes mushrooms, rusts, puffballs, etc.

Carrying capacity — The maximum level to which the resources of a habitat can be used without inducing instability.

City — An area where artifacts dominate the landscape and people are concentrated = Urban area = Built area = Urban environment.

Commensal — The relationship between two species which interact in such a way that one benefits while the other is unaffected.

Community — All the organisms living in one area.

Coppicing — Cutting of trees, down to stools, every fifteen years (e.g. hazel or chestnut).

Cover — Vegetation that serves to protect or conceal animals.

164

Cultivated area	— Land upon which flora and fauna are controlled to further human ends.
Ecology	— The study of relationships between organisms and their environment.
Ecotone	— A border zone between habitats (e.g. woodland edge, shoreline and riverbank).
Edge effect	— The tendency for increased variety and density at community junctions.
Edge species	— Organisms residing in "edge" habitat; these species often require more than one type of habitat in order to live and successfully reproduce.
Environment	— Everything outside the individual organism (used here in a loose way to mean physical surroundings).
Feral	— Animals which were once domesticated or possessed domesticated ancestors, but now living without support of a human owner.
Flora	— The sum of the plant species occupying an area at a given point in time.
Garden	— The open space associated with a house; includes lawns, flowerbeds, shrubbery, and vegetable patch = yard = backyard.
Green belt	— A planning device whereby undeveloped land surrounding an urban area is protected from unplanned development.
Green space	— Open space which has a vegetation cover = "green lung."
Habitat	— The place where an organism (or a community of organisms) lives. It is commonly defined with respect to the type of soil, topography and vegetation which occupy the area.
Indigenous	— Original, not subjected to man's influence = Primitive = Native.
Management	— Manipulation of natural materials and processes according to a predetermined plan.

Melanism — The condition of having a high amount of dark or black pigment.

Micro-environment — An environmental unit of small dimension (e.g., one meter–one millimeter) created by small differences in physical parameters.

Micro-habitat — A small unit of space in which a specific life form can exist and reproduce.

Natural area — An area where biotic and physical processes predominate over man's influence; thus a tract with its natural cover and associated fauna.

Open space — Land which is predominantly undeveloped = Unbuilt land.

Parkland — An area comprised of mature trees set in grassland.

Pollarding — The practice of lopping or trimming trees in order to produce a close-rounded head of young branches.

Pollution — Addition of foreign materials (e.g. pesticides) into the environment; disruption of energy fluxes and material cycles by human activities.

Propagule — Any part of a plant capable of forming a new organism = Fruit = Seed = Spore.

Riparian — Situated on the bank of a river or other body of water.

Semi-natural area — An area with flora and fauna that are native or introduced in the distant past, but where the vegetation has been modified by human activity so that it differs from the natural potential vegetation for the site.

Suburban habitat — Habitat within (or near) the city that has been extensively modified by man, usually dominated by organisms that appear in the early successional stages.

Trophic — Referring to the process of nutrition.

Urban environment — Cities and suburbs—wherever buildings, streets, gardens and other land uses have replaced the natural landscape; a contained, highly interrelated system of natural and man-made elements in various mixes.

Urbanization — The process in which rural depopulation is accompanied by a concentration of the populace in major urban centers.

Vegetation — Natural and semi-natural complexes of plant communities—the total plant cover of an area.

Wild — Nondomesticated, although not necessarily original or primitive.

Wildlife — All nondomesticated animals.

Wildlife corridors — Linear cover providing concealment, between open spaces or in built-up areas allowing dispersal and migration (e.g., hedges, highway verges).

Wild landscape — Natural area little changed by man's activities.

NOTE:

The two lists that follow include names of only a small portion of the organisms in North America and the British Isles. Our main purpose is to supply a useful list of scientific and common names for each of the plants and animals mentioned in this book. Most of the scientific names listed are genus and species names, though in several cases (e.g. ragweed, *Ambrosia* spp.) only the genus is given; and, in the case of bats, only Chiroptera, the order, is listed.

Although biologists recognize that scientific names of plants and animals are sometimes controversial (primarily because taxonomic schemes are continually subject to revision—and the revisions are seldom universally accepted by all professionals in the field), it is the common names that present the most problems. In our list, for example, we have "Mouse (*Micromys minutus*)"; but, of course, there are many other kinds of mice. The + preceding "Mouse" means that that particular mouse (*M. minutus*) is a British species. (Elsewhere in the list, notice the "House mouse" that is common to both North America and the British Isles.) Another problem with common names is that they vary greatly from place to place and from time to time. Many American readers, for example, may see 'Crow" with a + indicating that this bird does not occur in North America. Americans know, of course, that they often see crows; but the American crows are a different species and are listed as "Common crow" in this list. Despite such difficulties we hope that these two lists of the organisms mentioned in this book will be useful to many readers.

168

Common Names
and Scientific Equivalents

PLANTS

Apple	(*Malus* spp.)
+Beech	(*Fagus sylvaticus*)
Birch	(*Betula* spp.)
Blackthorn	(*Prunus spinosa*)
Bracken	(*Pteridium aquilinum*)
Bramble	(*Rubus* spp.)
*Chamise	(*Adenostema fasciculatum*)
Common Buckthorn	(*Rhamnus cathartica*)
+Common Primrose	(*Primula vulgaris*)
Elder	(*Sambucus nigra*)
Elm	(*Ulmus* spp.)
Gorse	(*Ulex* spp.)
Holly	(*Ilex* spp.)
Juniper	(*Juniperus communis*)
Larch	(*Larix* spp.)
Oak	(*Quercus* spp.)
Pine	(*Pinus* spp.)
Plane (=London plane tree)	(*Platanus acerifolia*)

*Indicates species that ordinarily occur in North America, but not in the British Isles.

+Indicates species that ordinarily occur in the British Isles, but not in North America.

Items printed without either of the above symbols are names of species that commonly occur both in the British Isles and in North America.

169

+Princess Tree   (*Paulownia* spp.)
*Ragweed   (*Ambrosia* spp.)
 Spruce   (*Picea* spp.)
 Sycamore maple   (*Acer pseudoplatanus*)
 Willow   (*Salix* spp.)

MAMMALS

+Badger   (*Meles meles*)
 Bat   (Chiroptera)
*Bear   (*Ursus* spp.)
*Black bear   (*Ursus americanus*)
 Black rat   (*Rattus rattus*)
*Common mole   (*Scalopus aquaticus*)
*Cougar   (*Felis concolor*)
*Coyote   (*Canis latrans*)
*Deer (=white-tailed
   deer)   (*Odocoileus virginianus*)
 Dog   (*Canis familiaris*)
 Domestic cat   (*Felis domestica*)
+English fox   (*Vulpes vulpes*)
+Fallow deer   (*Dama dama*)
 Fox   (*Vulpes* spp.)
 Grey squirrel   (*Sciurus carolinensis*)
 Hare   (*Lepus europaeus*)
+Hedgehog   (*Erinaceus europaeus*)
 House mouse   (*Mus musculus*)
*Lynx   (*Lynx canadensis*)
 Man   (*Homo sapiens*)
+Mole   (*Talpa europaea*)
+Mouse   (*Micromys minutus*)
 Norway rat   (*Rattus norvegicus*)
*Opossum   (*Didelphis marsupialis*)
+Rabbit   (*Oryctolagus cuniculus*)
*Raccoon   (*Procyon lotor*)
 Shrew   (*Sorex* spp.)
*Skunk   (*Spilogale putorius, Mephitis* spp).
 Stoat (=Ermine)   (*Mustela erminea*)
*Striped skunk   (*Mephitis mephitis*)
 Vole   (*Microtus* spp., *Clethrionomys* spp.)
+Weasel   (*Mustela nivalis*)
*Wolf   (*Canis lupus*)

BIRDS

*American robin	(*Turdus migratorius*)
*Barn owl	(*Tyto alba*)
*Barn swallow	(*Hirundo rustica*)
Black-backed gull	(*Larus fuscus*)
+Black Redstart	(*Phoenicurus ochruros*)
*Blue jay	(*Cyanocitta cristata*)
+Blue tit	(*Parus caeruleus*)
*Canada goose	(*Branta canadensis*)
*Chimney swift	(*Chaetura pelagica*)
*Common crow	(*Corvus brachyrhynchos*)
Coot	(*Fulica atra*)
+Crow	(*Corvus corone*)
+Dunnock	(*Prunella modularis*)
+English Blackbird	(*Turdus merula*)
+Great tit	(*Parus major*)
*Grouse	(*Lagopus lagopus*)
Gull	(*Larus* spp.)
+House martin	(*Delichon urbica*)
House sparrow	(*Passer domesticus*)
+Jackdaw	(*Corvus monedula*)
+Kestrel	(*Falco tinnunculus*)
+Kingfisher	(*Alcedo atthis*)
Magpie	(*Pica pica*)
Mallard	(*Anus platyrhynchos*)
*Mockingbird	(*Mimus polyglottos*)
Moorhen	(*Gallinula chloropus*)
Mute swan	(*Cygnus olor*)
Partridge	(*Perdix perdix*)
Peregrine falcon	(*Falco peregrinus*)
*Phoebe	(*Sayornis* spp.)
Pied wagtail	(*Motacilla alba*)
Pigeon (=Rock dove) (see also Wood pigeon)	(*Columba livia*)
*Pileated woodpecker	(*Dryocupus pileatus*)
*Red-headed woodpecker	(*Melanerpes erythrocephalus*)
+Robin	(*Erithacus rubecula*)
+Sand martin	(*Riparia riparia*)
+Song thrush	(*Turdus philomelos*)
Sparrow hawk	(*Falco sparverius*)
+Spotted woodpecker	(*Dendrocopus major*)
Starling	(*Sturnus vulgaris*)
+Swift	(*Apus apus*)
+Tawny owl	(*Strix aluco*)

Titmouse   (*Parus* spp.)
+Wood pigeon (=Ring dove)   (*Columba palumbus*)

AMPHIBIANS

Common frog   (*Rana temporaria*)
Toad   (*Bufo bufo*)

INSECTS

Bee   (*Apis* spp.)
Cockroach   (*Periplaneta* spp.)
+Peppered moth   (*Biston betularia*)

Scientific Names
and Common Equivalents

PLANTS

Acer pseudoplatanus	(Sycamore maple)
*Adenostema fasciculatum	(Chamise)
*Ambrosia spp.	(Ragweed)
Betula spp.	(Birch)
Fagus sylvaticus	(Beech)
Ilex spp.	(Holly)
Juniperus communis	(Juniper)
Larix spp.	(Larch)
Malus spp.	(Apple)
+Paulownia spp.	(Princess tree)
Picea spp.	(Spruce)
Pinus spp.	(Pine)
Platanus acerifolia	(Plane = London planetree)
+Primula vulgaris	(Common primrose)
Prunis spinosa	(Blackthorn)
Pteridium aquilinum	(Bracken)
Quercus spp.	(Oak)
Rhamnus cathartica	(Common buckthorn)
Rubus spp.	(Bramble)
Salix spp.	(Willow)
Sambucus nigra	(Elder)
Ulex spp.	(Gorse)
Ulmus spp.	(Elm)

MAMMALS

Canis familiaris	(Dog)
*Canis latrans	(Coyote)
*Canis lupus	(Wolf)
Chiroptera	(Bat)
+*Dama dama*	(Fallow deer)
**Didelphis marsupialis* (= *D. virginiana*)	(Opossum)
+*Erinaceus europaeus*	(Hedgehog)
*Felis concolor	(Cougar)
Felis domestica	(Domestic cat)
Homo sapiens	(Man)
Lepus europaeus	(Hare)
*Lynx canadensis	(Lynx)
+*Meles meles*	(Badger)
**Mephitis mephitis*	(Striped skunk)
+*Micromys minutus*	(Mouse)
Microtus spp., *Clethrionomys* spp.	(Vole)
Mus musculus	(House mouse)
Mustela erminea	(Stoat = Ermine)
+*Mustela nivalis*	(Weasel)
**Odocoileus virginianus*	(Deer = White-tailed deer)
+*Oryctolagus cuniculus*	(Rabbit)
*Procyon lotor	(Raccoon)
Rattus norvegicus	(Norway rat)
Rattus rattus	(Black rat)
**Scalopus aquaticus*	(Common mole)
Sciurus carolinensis	(Grey squirrel)
Sorex spp.	(Shrew)
**Spilogale putorius*, *Mephitis* spp.	(Skunk)
+*Talpa europaea*	(Mole)
*Ursus americanus	(Black bear)
*Ursus spp.	(Bear)
Vulpes spp.	(Fox)
+*Vulpes vulpes*	(English fox)

BIRDS

+*Alcedo atthis*	(Kingfisher)
Anus platyrhynchos	(Mallard)
+*Apus apus*	(Swift)
**Branta canadensis*	(Canada goose)

*Chaetura pelagica	(Chimney swift)
Columba livia	(Pigeon = Rock dove)
+Columba palumbus	(Wood pigeon)
*Corvus brachyrhynchos	(Common crow)
+Corvus corone	(Crow)
+Corvus monedula	(Jackdaw)
*Cyanocitta cristata	(Blue jay)
Cygnus olor	(Mute swan)
+Delichon urbica	(House martin)
+Dendrocopus major	(Spotted woodpecker)
*Dryocupus pileatus	(Pileated woodpecker)
+Erithacus rubecula	(Robin)
Falco sparverius	(Sparrowhawk)
+Falco tinnunculus	(Kestrel)
Falco peregrinus	(Peregrine falcon)
Fulica atra	(Coot)
Gallinula chloropus	(Moorhen)
*Hirundo rustica	(Barn swallow)
*Lagopus lagopus	(Grouse)
Larus fuscus	(Black-backed gull)
Larus spp.	(Gull)
*Melanerpes erythrocephalus	(Red-headed woodpecker)
*Minus polyglottos	(Mockingbird)
Motacilla alba	(Pied wagtail)
+Parus caeruleus	(Blue tit)
+Parus major	(Great tit)
Parus spp.	(Titmouse)
Passer domesticus	(House sparrow)
Perdix perdix	(Partridge)
+Phoenicurus ochruros	(Black redstart)
Pica pica	(Magpie)
+Prunella modularis	(Dunnock)
+Riparia riparia	(Sand martin)
*Sayornis spp.	(Phoebe)
+Strix aluco	(Tawny owl)
Sturnus vulgaris	(Starling)
+Turdus merula	(English blackbird)
*Turdus migratorius	(American robin)
+Turdus philomelos	(Song thrush)
*Tyto alba	(Barn owl)

AMPHIBIANS

Bufo bufo	(Toad)
Rana temporaria	(Common frog)

INSECTS

Apis spp.   (Bee)
+*Biston betularia*   (Peppered moth)
Periplaneta spp.   (Cockroach)

Bibliography

Abro, A. 1964. Fugl som apner melkeflasker. [Birds which open milk bottles.] *Sterna* 6(2):81−85. [Engl. summ.]

Air Pollution, 1969. Proceedings of the First European Congress on the Influence of Air Pollution on Plants and Animals. Wageningen, April 22−27, 1968. Centre for Agric. Publishing and Documentation, (pudoc) Wageningen, The Netherlands. 415 pp.

Allen, D. L. 1953. Wildlife habitat in relation to the use of herbicide sprays on farms, ranches and roadsides. *Proc. 43rd. Convention Int. Assoc. Game, Fish and Consn.,* Wisconsin: pp. 90−94.

Anderson, D. W., Hickey, J. J., Risebrough, R. W., Hughes, D. F., and Christensen, R. E. 1969. Significance of chlorinated hydrocarbon residues to breeding Pelicans and Cormorants. *Can. Field Nat.* 83:91−112.

Antonovics, J., and Bradshaw, A. D. 1970. Evolution in closely adjacent plant populations. VIII. Clinal patterns at a mine boundary. *Heredity* 25:349−362.

Appleton, J. H. 1970. *Disused Railways in the Countryside of England and Wales; a Report to the Countryside Commission.* HMSO.

Arvill, R. 1967. *Man and Environment; Crisis and the Strategy of Choice.* Penguin Books, London. 332 pp.

Aschmann, H. 1959. The evolution of wild landscape and its persistence in Southern California. *Annals,* Assoc. of Amer. Geogr. 49(3):34−56.

Atkinson-Willes, G. 1969. Wildfowl and recreation: a balance of requirements. *Brit. Water Supply* ll:5−15.

Augustine, M. T. 1966. Using vegetation to stabilize critical areas in building sites. *Soil Conservation* 32:78–79.

Ayres, R. U., and Kneese, A. V. 1969. *Pollution and environmental quality; The Quality of the Urban Environment,* ed. H. Perloff. Washington D.C.: Resources for the Future. pp. 35–71.

Balazova, G., and Hluchaň, E. 1969. [The effect of flourine exhalates on animals in the surroundings of an aluminum plant.] In *Air Pollution, 1969.* Wageningen. pp. 275–279.

Bangerter, E. B. 1961. The botany of the London area. *Lond. Nat.* 40:6–16.

Barnes, H. F., and Weil, J. W. 1944. Slugs in gardens: Their numbers, activities and distribution. Part I. *J. Anim. Ecology* 13:140–175.

———. 1945. Slugs in gardens: Their numbers, activities and distribution. Part II. *J. Anim. Ecology* 14:71–105.

Beames, I. R. 1968. Bats in the London area. *Lond. Nat.* 47:38–49.

Beck, A. M. 1973. *The Ecology of Stray Dogs.* Baltimore: York Press. 98 pp.

Beckwith, S. L. 1941. Preliminary study of succession on abandoned agricultural uplands and its relationships to wildlife management. Master's Thesis. Ann Arbor: University of Michigan.

———. 1950. Ecological succession on abandoned farm lands and its relationship to wildlife management. Ph. D. Thesis. Ann Arbor: University of Michigan.

Beecher, W. J. 1942. Nesting birds and the vegetative substrate. Chicago Orn. Soc., Chicago Mus. Nat. Hist. 69 pp.

Benthem, R. J. 1968. Creative conservation. *Biol. Conserv.* 1:11–12.

Benton, A. H., and Dickinson, L. 1966. Wires, poles, and birds. In *Birds in Our Lives,* ed. A. Stefferud. U. S. Dept. Interior. Washington, D. C. pp. 390–395.

Bergman, G. 1961. The food of birds of prey and owls in Fenno-Scandia, *Brit. Birds* 54:307–320.

Beven, G. 1964. The feeding sites of birds in grassland with thick scrub. *Lond. Nat.* 43:86–109.

———. 1965. The food of tawny owls in London. *Lond. Bird Rep.* 29:56–72.

Bishop, R. A., and Barratt, R. 1970. Use of artificial nest baskets by Mallards. *J. Wildl. Mgmt.* 34(4):734–738.

Blackett, A. 1970. Blue tits and gulls feeding by arc lights. *Brit. Birds* 63(3):136.

Boddington, H. P. 1971. Director of Parks and Cemeteries, London Borough of Merton. Personal Communication. Letter dated 6.1.71.

Boulding, K. E. 1966. The economics of the coming spaceship earth. In *Environmental Quality in a Growing Economy,* ed. H. Jarrett. Washington, D.C.: Resources for the Future. pp. 3–14.

Boyd, H. 1957. The use of hand-reared ducks for supplementing wild populations. *The Wildfowl Trust, 8th Ann. Rep:* 91–95.

Boyer, G. F., and Devitt, O. E. 1961. A significant increase in the birds of Luther Marsh, Ontario, following fresh-water impoundment. *Can. Field Nat.* 75(4):225–237.

Bradley, B. O., and Cook, A. H. 1951. Small marsh development in New York. *Trans. N. Amer. Wildl. Conf.* 16:251–264.

Bramble, W. C., and Byrnes, W. R. 1955. Effect of certain brush control techniques and materials on game food and cover on a power right-of-way. I. Penn. State Univ. Agr. Exp. Sta. Progress Rep. 126.

Brandt, C. S., and Heck, W. W. 1968. Effects of air pollutants on vegetation. In *Air Pollution,* ed. A. C. Stern. Vol. I. New York: Academic Press. pp. 401–433.

Brandwein, P. F. 1966. Origins of public policy and practice in conservation: early education and the conservation of sanative environments. In *Future Environments of North America,* eds. F. F. Darling and J. P. Milton. New York: Natural History Press. pp. 628–647.

Bridgman, C. J. 1962. Birds nesting in aircraft. *Brit. Birds* 55(11):461–470.

Brightman, F. H. 1959. Some factors influencing lichen growth in towns. *Lichenologist* 1:104–108.

Brooks, C. E. P. 1952. Selective annotated bibliography on urban climates. *Meteor. Abst. and Bibliog.* 3:734–773.

Brooks, P. 1970. Acquiring and protecting land. In *Challenge for Survival: Land, Air and Water for Man in Megalopolis,* ed. P. Dansereau. New York: Columbia U. Press. pp. 71–77.

Brown, E. P. 1963. The bird life of Holland Park; the effect of human interference. *Lond. Bird Rep.* 26:60–87.

———. 1964. The bird life of Holland Park, 1962-1963. *Lond. Bird Rep.* 28:69–78.

Brown, L. G., and Bellrose, F. C., Jr. 1943. Use of nesting boxes for wood ducks by other wildlife. *J. Wildl. Mgmt.* 7:298–306.

Brumsted, H. B., and Hewitt, O. H. 1952. Early investigations on artificial marsh development in New York. *Trans. N. Amer. Wildl. Conf.* 17:259–268.

Bruyns, M. 1961. De dichteid van de broedvogelbevolkingen in de bebouwde kommen. [The density of the population of birds nesting in residential areas.] *Levende Natur.* 64(9):193–199.

Bryant, S. 1970. [press release] October 5. Los Angeles Department of Animal Regulations. 4 pp. (mimeo.).

Bull, J. 1964. *Birds of the New York Area.* New York: Harper and Row.

Busse, P., and Olech, B. 1968. [On some problems of birds spending nights in nestboxes.] *Acta. Ornithol.* 11:1−26.

Butler, R. E. 1962. The buried rivers of London. *Lond. Nat.* 41:31−41.

Bye, A. E., Jr. 1969. The Bog; a New England garden that maintains itself. *Landscape Architecture* 60(1):29−31.

Cain, S. 1966. Biotope and habitat. In *Future Environments of North America,* eds. F. F. Darling and J. P. Milton. New York: Natural History Press. pp. 38−54.

Caldwell, Lynton K. 1966. Problems of applied ecology: perceptions, institutions, methods, and operational tools. *BioScience* 16:524−527.

Campbell, C. J. 1970. Ecological implications of riparian vegetation management. *J. Soil and Water Conserv.* 25(2):49−52.

Cannon, H. L., and Bowles, J. M. 1962. Contamination of vegetation by tetraethyl lead. *Science* 137:765−766.

Carlisle, A., and Brown, A. H. F. 1967. The influence of forest practices on woodland nature reserves. In *The Biotic Effects of Public Pressures on the Environments,* ed. E. Duffey. Monks Wood. Exp. Sta. (England), Symp. 3:69−81.

Carlson, C. W. and Menzies, J. D. 1971. Utilization of urban wastes in crop production. *BioScience* 21(12):561−564.

Chabreck, R. H. 1968. Weirs, plugs and artificial potholes for the management of wildlife in coastal marshes. In *Proc. of Marsh and Estuary Mgmt. Symp.,* ed. J. D. Newsom. Baton Rouge: Louisiana State Univ. 1967:178−192.

Chadwick, G. F. 1966. *The Park and the Town.* London: Architectural Press. 388 pp.

Chandler, T. J. 1965. *The Climate of London.* London: Hutchinson. 292 pp.

——. 1970. Mankind's impact on the atmosphere. *The Geogr. Mag.* 42(2):83−90.

Chapman, D. I. 1969. The fallow deer situation in Epping Forest, Essex, England. *Biol. Conserv.* 1(3):252−253.

Chow, T. J. 1970. Lead accumulation in roadside soil and grass. *Nature (Lond.)* 225:295−296.

Church, R. 1956. The Royal Parks of London. Min. of Works Guidebook. HMSO, London. 61 pp.

Civic Trust. 1966. Moving big trees. London.

———. 1967. Practice notes on transplanting semi-mature trees. London.

———. (n.d.) Derelict land; a study of industrial dereliction and how it may be redeemed. London. 70 pp.

Clawson, M. 1962. A positive approach to open space preservation. *J. Am. Inst. Planners* 28(2):124–129.

———. 1969. Open (uncovered) space as a new urban resource. In *The Quality of the Urban Environment,* ed. H. S. Perloff. Washington, D.C.: Resources for the Future, pp. 139–175.

Clayton, K. M. (ed.) 1964. *Guide to London excursions.* 20th Int. Geogr. Congress. London. 162 pp.

Coates, U. A. 1964. Experiment in grassland establishment on colliery shale, Bickershaw Reservoir Site, Abram, 1954-60. Lancs. County Council, County Planning Dept. England.

Cohen, E. 1967. Nestboxes. Brit. Trust for Ornithol. Field Guide 3. Tring, Herts. England. 46 pp.

Collinge, W. E. 1936. The food and feeding habits of the coot (*Fulica atra* Linn.). *Ibis* 6:35–39.

Collins, W. Gordon. 1970. *Proc. of Derelict Land Symposium.* Dept. of Civil Eng., Univ. of Leeds. Guilford, England: Iliffe Sci. and Tech. Publns. Ltd.

Colquhoun, M. K., and Morley, A. 1943. Vertical zonation in woodland bird communities. *J. Anim. Ecology* 12:75–81.

Colvin, B. 1970. *Land and landscape; evolution, design and control.* 2nd ed. London: John Murray. 412 pp.

Commins, B. T. and Walker, R. E. 1967. Observations from a ten year study of pollution at a site in the city of London. *Atmos. Environment* 1:49–68.

Commoner, B. 1963. *Science and Survival.* New York: Viking Press. 150 pp.

Cooke, A. S. 1970. The effect of p.p.'-DDT on tadpoles of the common frog (*Rana temporaria*). *Envir. Pollution* 1(1):57–71.

Coulson, J. C. 1961. The post-fledgling mortality of the blackbird in Great Britain. *Bird Study* 8(2):89–97.

Cox, P. 1970. The squirrels of Queen's Park and Philosopher's Walk. Dept. of Zoology, Univ. of Toronto. 40 pp. (mimeo.).

Craighead, J. J., and Stockstad, D. S. 1961. Evaluating the use of aerial nesting platforms by Canada geese. *J. Wildl. Mgmt.* 25(4):363-372.

Cramp, S. 1968. Town and country Woodpigeon. *Birds* 2(1):19–21.

Cramp, S. and Gooders, J. 1967. The return of the house martin. *London Bird Rep.* 31:93–98.

Crompton, C. T. 1966. The treatment of waste slate heaps. *Town, Country Planning Rev.* 37:291–304.

Croxton, W. C. 1928. Revegetation of Illinois coal stripped land. *Ecology* 9:155–175.

Cunningham, G. (ed.) 1964. *Day Tours.* Lqs Angeles Geogr. Soc., Publ. No. 1. Palo Alto, California: Pacific Books. 277 pp.

Czaja, J. 1968. (Birds in the area of airports and the methods of controlling them.) *Przegl. Zool.* 12(1):73–78. (Engl. Summ.)

Dagg, A. I. 1970. Wildlife in an urban area. *Naturaliste Can.* 97:201–212.

Dambach, C. A. 1944. A ten year ecological study of adjoining grazed and ungrazed woodlands in northeast Ohio. *Ecol. Monogr.* 14(3):255–270.

Dancer, W. S., and Hardy, A. V. 1969. *Greater London.* London: Cambridge Univ. Press. 85 pp.

Darby, N. C. 1951. The clearing of the English woodlands. *Geography* 36:71–83.

Darley, E. F. 1966. Studies on the effect of cement-kiln dust on vegetation. *J. Air Polln. Control Assoc.* 16:145–150.

———. 1969. The role of photochemical air pollution on vegetation. In *Air Pollution, 1969.* Wageningen: pp. 137–142.

Dasmann, R. F. 1964. *Wildlife Biology.* New York: J. Wiley & Sons. 231 p.

———. 1966. Wildlife in urban and suburban America. AAAS. 133rd Meeting. Washington, D.C. 8 p. (mimeo.).

———. 1968. *Environmental Conservation.* 2nd ed. New York: J. Wiley and Sons. 375 pp.

Davey, J. H. 1969. Nestboxes for birds of prey. *Birds* 2(7):174–175.

Davis, A. A. 1967. Metropolitan growth—encroachment or orderly progress. In *Man and the Quality of His Environment,* ed. J. E. Flack and M. C. Shipley. Boulder: U. Colorado Press. pp. 201–220.

Davis, B. N. K. 1967. Bird feeding preferences among different crops in an area near Huntingdon. *Bird Study* 14:227–237.

Davis, D. E. 1953. The characteristics of rat populations. *Quart. Rev. Biol.* 28:373–407.

Davison, V. E., and Grizzell, R. A. 1961. Choice food of birds—summer and fall. *Aud. Mag.* 63:162–167.

Davison, V. E., and Hamor. W. H. 1960. A system for classifying plant food of birds. *J. Wild. Mgmt.* 24:307–313.

Dennis, Eve (ed.). 1972. *Everyman's Nature Reserve: Ideas for Action.* Newton Abbot, Devon, England: David and Charles, Ltd. 256 pp.

van Dersal, W. R. 1941. Environmental improvement for valuable non-game animals. *N. Amer. Wildl. Conf.* 5:200–202.

Dhondt, A. A. 1970. The sex ratio of nestling great tits. *Bird Study* 17(3):282–286.

Dickinson, R. E. 1966. The process of urbanization. In *Future Environments of North America,* ed. F. F. Darling and J. P. Milton. New York: The Nat. Hist. Press. pp. 463–478.

Dobie, J. F. 1961. *The Voice of the Coyote.* Lincoln: U. of Nebraska Press. 386 pp.

Dowdy, W. W. 1947. An ecological study of the Arthropoda of an oak-hickory forest with reference to stratification. *Ecology* 28:418–439.

Downes, R. G. 1965. Conservation and the community, *J. Soil and Wat. Conserv.* 20(5):206–209.

Drury, W. H., Jr. 1966. Birds at airports. In *Birds in Our Lives,* ed. A. Stefferud. U. S. Dept. Interior. Washington, D.C. pp. 384–389.

Duffey, E. (ed.) 1967. *The Biotic Effects of Public Pressures on the Environment.* Monks Wood Exp. Sta. (England) Symp. 3 Nat. Conservancy, NERC, England. 178 pp.

Dunthorn, A. A., and Errington, F. P. 1964. Casualties among birds along a selected road in Wiltshire. *Bird Study* 11:168–182.

Dyrcz, A. 1969. The ecology of the song-thrush (*Turdus philomelos* Br.) and blackbird (*Turdus merula* L.) during the breeding season in an area of their joint occurrence. *Ecol. Polska* – Ser. A. 17(39):736–793.

Edwards, C. A. 1965. Effects of pesticide residues on soil invertebrates and plants. In *Ecology and the Industrial Society,* ed. G. T. Goodman *et al.* New York: J. Wiley and Sons, pp. 239–262.

Egler, F. E. 1952. Herbicide effects in Connecticut vegetation, 1950. *J. Forestry* 50(3):198–204.

———. 1954. Vegetation management for rights-of-way and roadsides. *Ann. Rep. Smithsonian Inst.* Publn. No. 4149. Washington, D.C. pp. 299–322.

Elton, C. S. 1958. *The Ecology of Invasions by Animals and Plants.* London: Methuen. 181 pp.

van Emden, H. F. 1965. The role of uncultivated land in the biology of crop pests and beneficial insects. *Sci. Hort.* 17:121–136.

Erz, W. 1959. [Population of birds in residential parts of a large industrial city in Westphalia.] *Ornithol. Mitteil.* 11(12):221−227.

———. 1964. [Influence of the severe winter 1962/3 on the population structure of urban blackbirds.] *Natur. und. Heimat* 24(4)84−85.

Fahey, J. E., and Butcher, J. W. 1965. Chlorinated hydrocarbon insecticide residues in soils of urban areas. Battle Creek, Michigan, *J. Econ. Ent.* 58(5):1026−1027.

Fairbrother, N. 1970. *New Lives, New Landscapes.* London: Architectural Press. 397 pp.

Fairhurst, A. R. 1970. Magpies nesting on ledges and on a crane. *Brit. Birds* 63:387.

Farness, S. 1966. Resources planning versus regional planning. In *Future Environments of North America,* eds. F. Fraser Darling and J. P. Milton. New York: Natural History Press. pp. 494−502.

Felton, P. M. and Lull, H. W. 1963. Suburban hydrology can improve watershed conditions. *Public Works* 94:93−94.

Fenton, A. F. 1960. Lichens as indicators of atmospheric pollution. *Irish Nat. J.* 13:153−159.

Finnis, R. G. 1960. Road casualties among birds. *Bird Study* 7(1):21−32.

Fischer, R. B. 1958. The breeding biology of the Chimney Swift (*Chaetura pelagica L.*). New York State Mus. and Sci. Serv. Bull. No. 368.

Fisher, J., and Hinde, R. A. 1949. The opening of milk bottles by birds. *Brit. Birds.* 32:347−357.

Fitter, R. S. R. 1945. *London's Natural History.* London: Collins. 282 pp.

———. 1963. *Wildlife in Britain.* Harmondsworth, Middlesex, England: Penguin Books. 191 pp.

Floyd, C. 1965. Conservation of chalk grassland. *Soc. for Promotion of Nat. Reserves Hdbk. 48th Ann. Rep.* (British Museum [Nat. Hist.]). pp. 59−61.

Forman, B. 1968. Recreation in the countryside; impacts and opportunities. *SPNR Hdbk.* 1968: 28−35.

Fosberg, F. R. 1966. Restoration of lost and degraded habitats. In *Future Environments of North America,* ed. F. F. Darling and J. P. Milton. New York: Natural History Press. pp. 503−515.

Fowles, J. 1970. Weeds, bugs, Americans. *Sports Illustrated* 33:86−102.

Froman, R. 1961. *The Nerve of Some Animals.* Philadelphia: J. B. Lippincott Co. pp. 108−127.

Funk, D. T. 1962. A revised bibliography of strip-mine reclamation. USDA. Forest Service. Central States Forest Exp. Sta. Columbus, Ohio, Misc. Rep. 35.

Galbraith, J. K. 1967. *The New Industrial State.* New York: New Amer. Library. 430 pp.

Geiger, R. 1965. *The Climate Near the Ground.* Cambridge, Massachusetts: Harvard U. Press, 611 pp.

Gibbs, A. 1963. The bird population of rubbish dumps. *Lond. Bird Rep.* 26:104⁻110.

Gilbert, O. L. 1965. Lichens as indicators of air pollution in the Tyne Valley. In *Ecology and the Industrial Society,* eds. G. T. Goodman *et al.* New York: J. Wiley and Sons. pp. 35⁻47.

———. 1969. The effect of sulfur dioxide on lichens and bryophytes around Newcastle-upon-Tyne. In *Air Pollution, 1969.* Wageningen. pp. 223⁻235.

———. 1970. A biological scale for the estimation of sulfur dioxide pollution. *New Phytol.* 69(3) 629⁻634.

Giles, R. H. 1969. *Wildlife Management Techniques.* 3rd ed. rev. Washington, D.C.: The Wildlife Soc. 623 pp.

Gill, D. 1965. Coyote and urban man: a geographic analysis of the relationship between coyote and man in Los Angeles. Master's Thesis, Dept. of Geography, University of California at Los Angeles, 114 pp.

———. 1966. Coyote and man in Los Angeles. *British Columbia Geogr. Series* 7:69⁻84.

———. 1970. The coyote and the sequential occupants of the Los Angeles basin. *Amer. Anthropologist* 72:821⁻826.

Gladwin, T. W. 1963. A short account of Rye Meads, Herts, and its ornithology. *Lond. Bird Rep.* 26:88⁻99.

Glue, D. E. 1970. Changes in the bird community of a Hampshire gravel pit, 1963-1968. *Bird Study* 17(1):15⁻27.

Gompertz, T. 1957. Some observations on the feral pigeon in London. *Bird Study* 4:2⁻13.

Gooders, J. 1965. The birds of Clapham and Wandsworth Commons. *Lond. Bird Rep.* 29:73⁻88.

———. 1968. The swift of central London. *Lond. Bird Rep. 32:93⁻*98.

Goodman, G. T., Edwards, R. W., and Lambert, J. L. (ed.) 1965. *Ecology and the Industrial Society.* Brit. Ecol. Soc. Symp. 5. New York: J. Wiley and Sons, 395 pp.

Goodman, W. I. and Freund E. C. (ed.) 1968. *Principles and Practice of Urban Planning.* 4th edn. Washington D.C.: Intl. City Managers Assoc. 621 pp.

Goodwin, D. 1954. Notes on feral pigeons. *Avicult. Mag.* 60:190–213.

Goodwin, R. H., and Niering, W. A. 1959. The management of roadside vegetation by selective herbicide techniques. *Connecticut Arboretum Bull.* 11:4–10.

Gottmann, J. 1970. The green areas of megalopolis. In *Challenge for Survival; Land, Air, and Water for Man in Megalopolis,* ed. P. Dansereau. New York: Columbia U. Press. pp. 61–65.

Gottschalk, J. S. 1967. The introduction of exotic animals into the United States. Intern'l. Union Conserv. Nature and Nat. Res. 10th Tech. Meeting, Lucerne. pp. 113–119.

Graham, E. H. 1947. *The Land and Wildlife.* New York: Oxford Univ. Press. 232 pp.

Grant, P. J. 1967. The birds of Greenwich Park and Blackheath. *Lond. Bird Rep.* 31:64–92.

———. 1970. Duck on the River Thames at Woolwich. *Lond. Bird Rep.* 34:80–85.

Green, R. C., and Forsyth, E. M. 1970. Regent's Park and Primrose Hill, in *Bird Life in the Royal Parks, 1967–1968.* Report by the Committee on Bird Sanctuaries in the Royal Parks. Min. Pub. Bldg. and Works. HMSO. London. pp. 9–10.

Groskin, H. 1952. Observations of duck hawks nesting on man-made structures. *Auk* 69:246–253.

Grummt, W. 1962. [Successful winter breeding of the Blackbird (*Turdus merula*) in Berlin.] *Vogelwarte* 21(4):295–296.

Guichard, G. 1957. Le régime alimentaire parisien de la hulotte (*Strix aluco sylvatica* Shaw.) *Oiseau* 27:140–142.

Guy, H. P., and Ferguson, G. E. 1970. Stream sediment: an environmental problem. *J. Soil and Water Conserv.* 25(6):217–221.

Hackett, B. 1969. Ecology and technology in twentieth-century landscapes. *Biol. Conserv.* 1(2):117–122.

Haddock, P. 1961. Coniferous shade trees in suburban development. *in 28th West. Chapter Nat. Shade Tree Conf. Vancouver, Canada. 1961. Nat. Shade Tree Conf. Proc.* 37:147–159.

Hall, G. H. 1970. The story of the Sun Life falcons. *Can. Field Nat.* 84(3):210–230.

Hancock, B. D. 1963. Some observations on bats in East Surrey and recent records for the London area. *Lond. Nat.* 42:26–41.

Hansen, R. P. 1968. Urban open space: psycho-socio and legal dimensions. In *Man and the Quality of His Environment,* eds. J. E. Flack and M. C. Shipley. Western Resources Conf. 1967. Boulder, Colorado: Univ. of Colorado Press. pp. 209–220.

Harper, J. F., and Morris, P. A. 1965. The occurrence of small mammals in discarded bottles. *Proc. Zool. Lond.* 145(1):148–153.

Harrison, C. J. O. 1960. The food of some urban tawny owls. *Bird Study* 7(4):236–240.

Harrison, J. 1963. Heavy mortality of mute swan from electrocution. *The Wildfowl Trust, 14th Ann. Rep.* pp. 164–165.

Harrison, J. G. 1970. Creating a wetland habitat. *Bird Study* 17(2):111–122.

Harrison, J., and Harrison, J. 1964. The management of a gravel pit wildfowl reserve. *Trans. VIth Congress. Int. Union of Game Biol.* London: Nat. Conservancy. pp. 323–329.

Hart, S. A. 1968. Composting: European activity and American potential. U.S. Pub. Health Service Pub. No. 1826.

Hartley, P. H. T. 1954. Wild fruits in the diet of British thrushes; a study in the ecology of closely allied species. *Brit. Birds* 47(4):97–107.

——. 1966. The bird garden. Royal Society for the Protection of Birds. Sandy, Beds. England. 16 pp.

Havlin, J. 1963. Rozmnozovani kosa cerneho (*Turdus merula*). [Reproduction in the Blackbird.] *Zool. Listy* 12(3):195–216. [Engl. Summ.]

Hawkins, A. F. 1970. Incubating moorhen repeatedly pulling cover over itself in rain. *Brit. Birds* 63(1):33–34.

Hay, K. 1969. An urban program for the Bureau of Sport Fisheries and Wildlife. Bur. Sport Fisheries and Wildlife. U.S. Dept. Interior. Washington, D.C. 44 pp. (mimeo.)

Hebblethwaite, R. L. 1969. The conservation of trees and shrubs in built-up areas. *Suppl. 3. J. Devon Trust for Nature Conserv. 4* (DTNC, Exeter.)

Heyl, D. H. 1954. Preliminary investigations of ecological requirements of shrubs on abandoned land. Master's Thesis. Penn. State U., University Park, Pennsylvania.

Hickey, J. J. 1966. Birds and pesticides. In *Birds in Our Lives*, ed. A. Stefferud. U.S. Dept. Interior. Washington, D.C. pp. 318–329.

——. 1970. Peregrine falcons, pollutants, and propaganda. *Can. Field Nat.* 84:207–208.

Hickey, J. J., and Hunt, L. B. 1960. Initial songbird mortality following a Dutch elm disease control program. *J. Wildl. Mgmt.* 24(3):259–265.

Hochbaum, H. A. 1965. The Wascana Waterfowl Park. Wascana Centre Authority. Regina, Sask. 64 pp.

Hodson, N. L. 1960. A survey of vertebrate road mortality. *Bird Study* 7:224−231.

——. 1962. Some notes on the causes of bird road casualties. *Bird Study* 9(3):168−173.

Hoffman, R. H. 1970. Waterfowl utilization of ponds blasted at Delta, Manitoba. *J. Wildl. Mgmt.* 34(3):586−593.

Holliday, R., Townsend, W. N., Hodgson, D. R., and Wood, J. W. 1958. Plant growth on "fly ash." *Nature (Lond.)* 176:1079−1080.

——. 1961. Restoration of pulverised fuel ash covered land. Leeds University, England.

Hooper, M. D. 1970. Disappearing hedgerows in the United Kingdom: the effect on conservation. *Biol. Conserv.* 2(3):230−231.

Hooper, M. D., and Holdgate, M. W. (ed.) 1970. Hedges and hedgerow trees. Monks Wood Symp. 3. Nat. Cons., Monks Wood, England. 104 pp.

Hooper, R. G., and Crawford, H. S. 1969. Woodland habitat research for non-game birds. *Trans. N. Amer. Wildl. Conf.* 34:201−207.

Hopkins, W. L. 1961. Chemical roadside weed control. *Proc. 13th Ann. Calif. Weed Conf.* pp. 15−22.

Howard, W. E. 1965. Interaction of behaviour ecology and genetics of introduced mammals. In *The Genetics of Colonizing Species,* eds. H. G. Baker and G. L. Stebbins. Proc. First Int. Union of Biol. Sci. Symp. on General Biology. New York: Academic Press. pp. 461−484.

——. 1967. Ecological changes in New Zealand due to introduced mammals. IUCN 10th Tech. Meeting, Lucerne. 1966. pp. 219−240.

Howard, W. E. 1969. The population crisis is here now. *Bio-Science* 19:779−784.

Hughes, D. L. 1957. Man and his animals. *Vet. Rec.* 69:1061−1065.

Hurcomb, Lord. 1969. Protection of wildlife in London and its outskirts by public authorities. *Biol. Conserv.* 1(2):166−169.

Hynes, H. B. 1960. *The Biology of Polluted Waters.* Liverpool: Univ. Press. 202 pp.

——. 1965. A survey of water pollution problems. In *Ecology and the Industrial Society,* eds. G. T. Goodman *et al.* New York: J. Wiley and Sons. pp. 49−63.

Ibberson, J. E., and Egler, F. E. 1951. Right-of-way maintenance by the selective application of selective herbicides. *Penn. For. and Water* 3(6):114−115, 125.

Ing, B. 1964. An unusual Myxomycete in the London area. *Lond. Nat.* 43:16−17.

Ingles, L. G. 1954. *Mammals of California and Its Coastal Waters.* Stanford, California: Stanford Univ. Press. 304 pp.

Jackson, W. B. 1951. Food habits of Baltimore, Maryland cats in relation to rat population. *J. Mammal.* 32:458–461.

Jacobson, A. R. 1968. Viable particles in the air. In *Air Pollution,* ed. A. C. Stern. Vol. I. pp. 95–119. New York: Academic Press.

Jaeger, E. 1950. The coyote as a seed distributor. *J. Mammal.* 31(4):350–359.

Jaffa, G. 1970. Clay, trees and planning. *T. C. Planning* 38(3):162–164.

James, N. D. G. 1939. The planting of tips and slag heaps. *Quart. J. Forestry* 33:164–172.

Jellicoe, G. A. 1966. *Studies in Landscape Design.* Vol. II. London: Oxford U. Press. 126 pp.

———. 1970. *Studies in Landscape Design.* Vol. III. Oxford Univ. Press, London 121 pp.

Jenkins, I. 1969. Increase in average of sunshine in Greater London. *Weather* 24:52–54.

Jensen, D. R. 1967. Selective land use for sand and gravel sites. Natl. Sand and Gravel Assoc. and Univ. of Illinois. 66 pp.

Johnson, W. 1930. *Animal Life in London.* London: The Sheldon Press. 171 pp.

Johnston, D. W., and Odum, E. P. 1956. Breeding bird populations in relation to plant succession on the Piedmont of Georgia. *Ecology* 37:50–62.

Jones, A. W. 1961. The vegetation of the South Norwood or Elmers End sewage works. *Lond. Nat.* 40:102–114.

Jones, J. C. 1940. Food habits of the American coot with notes on distribution. USDI. Bur. Biol. Survey. Wildl. Res. Bull. 2. 52 pp.

Kalmbach, E. R., and McAtee, W. L. 1969. Homes for birds. Consn, Bull. 14. USDI. BSFW. Washington D.C. 18 pp.

Karr, J. R. 1968. Habitat and avian diversity on strip-mined land in E. Central Illinois. *Condor* 70(4):348–357.

Karstad, L. 1967. Fluorosis in Deer (*Odocoileus virginianus*). *Bull. Wildl. Disease Assoc.* 3(2):42–46.

Katz, M. 1967. Effects of contaminants, other than sulphur dioxide, on vegetation and animals. In Can. Cl. Resource Min. *Pollution and our environment.* Conf. Proc. 1967. I:A4-2-2(1-18).

Kendeigh, S. C., and West, G. C. 1965. Caloric values of plant seeds eaten by birds. *Ecology* 46(4):553–555.

Kenfield, W. G. 1966. *The Wild Gardener in the Wild Landscape.* Hafner, New York. 232 pp.

Kent, D. H. 1961. The flora of Middlesex walls. *Lond. Nat.* 40:29–43.

Kent, T. J. 1970. Open space for the San Francisco area: organizing to guide metropolitan growth. Berkeley: Inst. of Governmental Studies, U. of California. 85 pp.

Kettlewell, H. B. D. 1958. A survey of the frequencies of (*Biston betularia* L.) and its melanic forms in Great Britain. *Heredity* 12:51–72.

———. 1961. The phenomenon of industrial melanism in Lepidoptera. *Ann. Rev. Ent.* 6:245–262.

Kieran, J. F. 1972. *A Natural History of New York City.* Boston: Houghton Mifflin. 428 pp.

Kirkpatrick, C. M. 1971. Urban ecology today. *The Wildl. Soc. News* 133:13–14.

Kloke, A., and Leh, H.-O. 1969. Verunreinigungen von kulturpflanzen mit blei ans kraftfahrzeugabgasen. [Pollution of cultivated plants with lead from auto exhaust.] In *Air Pollution,* 1969. Wageningen. pp. 259–268.

Klopfer, P. H. 1969. *Habitats and Territories: A Study of the Use of Space by Animals.* New York: Basic Books. 117 pp.

Knabe, W. 1965. Observations on world-wide efforts to reclaim industrial waste land. In *Ecology and the Industrial Society,* eds. G. T. Goodman *et al.* New York: J. Wiley and Sons, pp. 263–296.

Knechtel, A. 1909. Shade trees for prairie cities. *Can. Forestry J.* 5(2):84–88.

Koller, P., and Knabe, W. 1962. The reclamation of lands stripped for brown coal. Ohio Agr. Exp. Sta. For. Dept. Series 49.

Komarek, E. V., and Wright, E. G. 1929. Bird casualties on highways. *Wilson Bull.* 41:106.

Krull, J. N. 1970. Aquatic plant-macroinvertebrate associations and waterfowl. *J. Wildl. Mgmt.* 34(4):707–718.

Kubichek, W. F. 1941. Collecting and storing seeds of waterfowl food plants for propagation. *Trans. N. Amer. Wildl. Conf.* 5:364–368.

Lack, D. 1944. Blackbird spring song. *Brit. Birds* 38:116.

———. 1965. *The Life of the Robin.* 4th ed. London: Witherby. 240 pp.

Lagerwerff, J. V., and Specht, A. W. 1970. Contamination of roadside soil and vegetation with cadmium, nickel, lead and zinc. *Environ. Sci. and Technol.* 4(7):583–586.

Lancs. Naturalists' Trust. 1970. Annual Report. Birkenhead, Cheshire. England.

Landsberg, H. E. 1956. The climate of towns, in *Man's Role in Changing the Face of the Earth,* ed. W. L. Thomas, Chicago: U. of Chicago Press. pp. 584–606.

———. 1962. *Physical Climatology.* 2nd ed. Dubois, Pa: Gray Printing Co. 446 pp.

Larimer, E. J. 1969. A concept plan for the Middle Patuxent River Valley as quality environment. Bur. of Sport Fisheries and Wildlife, U.S. Dept. Interior. Atlanta, Georgia. 15 pp. (mimeo.)

Laundon, J. R. 1970. London's lichens. *Lond. Nat.* 49:20–69.

Lawrence, W. H. 1967. Effects of vegetation management on wildlife. In *Herbicides and Vegetation Management,* ed. M. Newton. Proc. symp. Corvallis: Oregon State University. pp. 88–93.

LeBlanc, F., 1961. Influence de l'atmosphère poluée des grandes aggomérations urbaines sur les épiphytés cortiocoles. *Rev. Can. Biol.* 20:823–827.

Leefe, J. D. 1968. Tree planting on the farm. Min. Agr. Fish. and Food. (Nat. Agr. Adv. Service) East Midland Region, Lindsey. 14 pp.

Leigh, E. G. 1965. On the relation between the productivity, biomass, diversity, and stability of a community. *Proc. Nat. Acad. Sci.* 53:777–783.

Leonard, R. E., and Parr, S. B. 1970. Trees as a sound barrier. *J. of Forestry* 68(5):282–283.

Leopold, A. 1933. *Game Management.* New York: Scribner's. 481 pp.

Leopold, A. S. et al. 1964. Predator and rodent control in the United States. Advisory Board on Wildlife Management, U.S. Dept. Interior. Washington, D.C. 14 pp. (mimeo.).

Lepine, M. P., and Sautter, V. 1951. Sur l'infection des pigeons parisiennes par le virus de l'orinthose. *Bull. Acad. Natn. Med.* 135:332–338.

Lewin, V. 1971. Exotic game birds of the Puu Waawaa Ranch, Hawaii. *J. Wildl. Mgmt.* 35(1):141–155.

Lihnell, D. 1969. Sulphate contents of tree leaves as an indicator of sulfur dioxide air pollution in industrial areas. In *Air Pollution, 1969.* Wageningen. pp. 341–352.

Limstrom, G. 1953. A bibliography of strip-mine reclamation. Central States For. Exp. Sta. Misc. Release 8.

Lindsey County Council. 1970. *Countryside Recreation: The Ecological Implications.* Lindsey, Lincs. England. 125 pp.

Little, C. E. 1968. *Challenge of the Land.* New York: Open Space Action Institute. 151 pp.

Ljunggren, L. 1968. Season studies of wood pigeon populations. *Viltrevy* 5:434–504.

Lloyd, S. 1944. Blackbirds singing in early spring in central London. *Brit. Birds* 38:216.

Longenecker, G. W., and Ellarson, R. 1960. Landscape plants that attract birds. Univ. of Wisconsin. Ext. Service Circ. 514. 10 pp.

Los Angeles Department of Animal Regulations, Memo, February 11, 1959.

Los Angeles Times, June 5, 1936; March 8, 1937; July 14, 1938; August 13, 1938; March 25, 1943; March 7, 1950; September 1, 1955; September 23, 1960; September 29, 1960.

Lowry, N. P. 1967. The climate of cities. *Sci. Amer.* 217(2):15–23.

MacAllister, E. 1944. Blackbirds singing in London in January. *Brit. Birds* 38:194.

MacArthur, R. H. 1964. Environmental factors affecting bird species diversity. *Amer. Nat.* 98(903):387–397.

MacArthur, R. H., MacArthur, J. W., and Preer, J. 1962. On bird species diversity. II. Prediction of birds census from habitat measurements. *Amer. Nat.* 96:167–174.

McCann, J. A. 1960. An analysis of factors influencing road kill of pheasants in the Willamette Valley. Master's Thesis. Corvallis, Oregon: Oregon State University.

McClintock, D. 1964. Natural history of the garden of Buckingham Palace: wild and naturalized vascular plants. *Proc. Trans. S. Lond. Ent. Soc.* 1963(2):14–25.

McHarg, I. L. 1966. Ecological determinism. In *Future Environments of North America,* eds. F. F. Darling and J. P. Milton, New York: Natural History Press. pp. 526–538.

——. 1969. *Design with Nature.* New York: Natural History Press, 197 pp.

Mackenthun, K. M. 1969. *The Practice of Water Pollution Biology.* U.S. Dept. Interior, Fed. Water Polln. Control Agency. Washington, D.C. 281 pp.

McKnight, T. 1964. *Feral Livestock in Anglo-America.* Berkeley: U. of Calif. Press. 87 pp.

McNab, B. K. 1963. Bioenergetics and the determination of home range size. *Amer. Nat.* 97:133–140.

Mandelker, D. R. 1962. *Green Belts and Urban growth.* Madison: U. Wisconsin Press. 176 pp.

Mann, L. 1970. Urban open space programming. *J. Am. Inst. of Planners* 36(1):65–68.

Manns, L. 1967. Leaf litter fauna in the ecology of ground feeding birds. *Lond. Nat.* 46:116–125.

Margalef, R. 1968. *Perspectives in Ecological Theory.* Chicago: U. of Chicago Press. 111 pp.

Marks, J. B. 1942. Land use and plant succession in Coon Valley, Wisconsin. *Ecol. Monogr.* 12(2):113–133.

Marlborough, D. 1963. A supplement to "the fishes of the London area." *Lond. Nat.* 42:62–70.

Marling, R. J. 1963. Trees including preservation, planting, law and highways. *Estates Gazette Ltd.* (London).

Martin, A. C., Zim, H. S., and Nelson, A. L. 1951. *American Wildlife and Plants.* New York: McGraw-Hill Co. 500 pp.

Mason, G. 1965. Wildlife Problems and Control in the City of Los Angeles. Unpublished Report, District Supervisor, Los Angeles Department of Animal Regulations. 14 pp.

Maunder, W. J. (ed.) 1969. *Pollution.* Victoria, B.C.: U. of Victoria. 115 pp.

Meadows, B. S. 1961. The gull roosts of the Lea Valley reservoirs. *Lond. Bird Rep.* 25:56–60.

——. 1965. Black Redstarts along the lower reaches of the River Lea. *Lond. Bird Rep.* 29:91–94.

——. 1970. Breeding distribution and feeding ecology of the Black Redstart in London. *Lond. Bird Rep.* 34:72–79.

Mellanby, K. 1967. *Pesticides and Pollution.* London: Collins. 221 pp.

——. 1968. The effects of some mammals and birds on regeneration of oak. *J. Appl. Ecology* 5:359–366.

Meyer, J. G. 1972. Renewing the soil. *Environment* 14(2):22–32.

Meyer, K. F. 1959. Some general remarks and new observations on psittacosis and ornithosis. *Bull. World Hlth. Org.* 20:101–119.

Miller, H. W. 1969. Wildlife on man-made water areas. Notes from a seminar. U.S. Dept. Interior, BSFW. N. Prairie Wildl. Res. Center. Jamestown, N. Dakota. 28 pp. (mimeo.).

Minckler, L. S. 1946. Old field reforestation in the Great Appalachian Valley as related to some ecological factors. *Ecol. Monogr.* 16(1):87–108.

Ministry of Housing and Local Government. 1958. Trees in town and city. HMSO, London. 83 pp.

——. 1963. New life for dead lands. HMSO, London 30 pp.

Ministry of Public Building and Works. 1970. Bird life in the Royal Parks, 1967–1968. Rep. by Committee on Bird Sanctuaries in the Royal Parks. HMSO, London. 34 pp.

Mitcham Common Survey. 1968. London Borough of Merton, Morden. Surrey. England. 8 pp. (mimeo.).

Mitchell, K. D. G. 1967. Nocturnal activity of City Blackbird. *Brit. Birds.* 60(9):373–374.

Montier, D. 1968. A survey of the breeding distribution of the

Kestrel, Barn Owl and Tawny Owl in the London area in 1967. *Lond. Bird Rep.* 32:81−92.

Moore, N. W. 1962. The heaths of Dorset and their conservation. *J. Ecology* 50:369−391.

——. 1965. Environmental contamination by pesticides. In *Ecology and the Industrial Society*, eds. G. T. Goodman *et al.* New York: J. Wiley and Sons. pp. 35−47.

Morrow, P. E. 1964. Animals in toxic environments; mammals in polluted air. *In Handbook of Physiology*, ed. D. B. Dill, Sect. 4. Washington, D.C.: Amer. Physiol. Soc. pp. 795−808.

Munro, D. A., and Harris, R. D. 1963. *Du danger que constituent les oiseaux près des aérodromes du Canada.* Colloque le problême des oiseaux sur les aérodromes. Paris: Natl. de la Rech. Agroniomique. pp. 173-206.

Murton, R. K. 1965. Natural and artificial population control in the wood pigeon. *Ann. appl. Biol.* 55:177-192.

Murton, R. K., and Westwood, N. J. 1966. The foods of the Rock Dove and Feral Pigeon. *Bird Study* 13(2):130−146.

Murton, R. K., and Wright, E. N. (ed.) 1968. *The Problems of Birds as Pests.* Proc. of Inst. Biol. Symp. 17. London: Academic Press. 254 pp.

Newman, L. H. 1966. *Man and Insects: Insect Allies and Enemies.* New York: Natural History Press. 252 pp.

Newton, M. (ed.) 1967. *Herbicides and Vegetation Management.* Proc. Symp. Corvallis: Oregon State U. 356 pp.

Nicholson, E. M. 1956. Nature conservation and the management of natural area. IUCN. 6th Tech. Meeting, Edinburgh. pp. 8−26.

Niering, W. A. 1960. Nature in the metropolis. Park, Recn. and Open Space Project of Regl. Plan. Assoc., New York. No. 528.

——. 1968. The ecology of wetlands in urban areas. *Gardens Journal* 18(6):177−183.

——. 1969. Pesticides in suburbia. In *Current Topics in Plant Science*, ed. J. E. Gunckel, New York: Acad. Press. pp. 403−419.

——. and Goodwin, R. H. 1963. *Creating New Landscapes with Herbicides.* Connecticut Arboretum Bull. 14.

Nováková, E. 1969. Influence des pollutions industrielles sur les communautés animales et l'utilisation des animaux comme bioindicateurs. In *Air Pollution, 1969.* Wageningen. pp. 41−48.

Odum, E. P. 1959. *Fundamentals of Ecology.* 2nd ed. Philadelphia: W. B. Saunders Co. 546 pp.

——. 1969. The strategy of ecosystem development. *Science* 164(3877):262−270.

Oetting, R., and Cassel, J. F. 1971. Waterfowl nesting on interstate highway right-of-way in North Dakota. *J. Wildlife Mgt.* 35(4):774−781.

Olney, P. J. S. 1958. Food and feeding habits of wildfowl. The Wildfowl Trust, 9th Ann. Rep. pp. 47−51.

———. 1964. Gravel pits as waterfowl reserves. *Proc. MAR Conf. Intern'l. Union Conserv. Nature and Nat. Res. Publn.* 3 Vol. I, pp. 414−20.

———. 1967. The feeding ecology of local mallard and other waterfowl. *The Wildfowl Trust. 18th Ann. Rep.* pp. 47−55.

———. 1968. The food and feeding habits of the pochard. *Biol. Conserv.* 1(1):71−76.

Otto, H. W., and Daines, R. H. 1969. Plant injury by air pollutants: Influence of humidity on stomatal aperatures and plant response to ozone. *Science* 163(3872):1209−1210.

Owen, B. A. 1967. Lesser black-backed gulls nesting on warehouse roofs inland. *Brit. Birds* 60:416.

Owen, D. F. 1956. The food of nestling jays and magpies. *Bird Study* 3(4):257−265.

Parmenter, L. 1968. More flies of the Cripplegate bombed site, City of London. *Lond. Nat.* 47:81−86.

Parr, D. 1963. Bird life on a sewage disposal works. *Lond. Bird Rep.* 27:66−90.

Pasadena Star-News, March 17, 1965; February 6, 1970; February 7, 1970.

Peal, R. E. F. 1965. Woodpigeons in a London suburb. *Lond. Bird Rep.* 29:89−90.

Pearson, L., and Skye, E. 1965. Air pollution affects pattern of photosynthesis in Parmelia sulcata, a corticolous lichen. *Science* 148:1600−1602.

Perring, F. H. 1967a. Changes in chalk grassland caused by galloping. *The Biotic Effects of Public Pressures on the Environment*, ed. E. Duffey. Monks Wood Exp. Sta. (England), Symp. 3. pp. 134−142.

———. 1967b. Verges are vital—a botanist looks at our roadsides. *J. Inst. Highway Eng.* 14:13−16.

Perrins, C. 1968. The great tit and the blue tit. *Birds* 2(2):32−33

Persson, B. 1968. Nagot om tornsangarens (*Sylvia communis*) hackning resultat i biocidbehandlad miljo. [About the breeding result of the whitethroat in habitats treated with biocides.] *Var Fagelvarld* 27(3):231−243. [Engl. Summ.]

Peterken, J. H. G. 1953. Habitats of the London area. *Lond. Nat.* 32:2−12.

Peters, H. 1948. Ergebnisse der Rattenbekämpfung in Stuttgart, 1944−48. *Festschrift zur Tag. Hyg. und Mikrobio.* Stuttgart. pp. 1−48.

Peters, T. H. 1970. Using vegetation to stabilize mine tailings. *J. Soil. Water Conserv.* 25(2):65−66.

Pickles, W. 1942. Animal mortality on three miles of Yorkshire roads. *J. Animal Ecology* 11(1):37−43.

Picozzi, N. 1968. Grouse bags in relation to the management and geology of heather moors. *J. Appl. Ecology* 5:483−488.

Pimlott, D. H. 1969. The value of diversity. *Trans. N. Amer. Wildf. Conf.* 34:265−280.

Platt, J. 1969. What we must do. *Science* 166:1115−1121.

Pollard, D. F. W. 1967. An appraisal of the planting program, 1959−66. *The Wildfowl Trust, 18th Ann. Rep.* pp. 55−62.

Porterfield, N. 1969. Ecological basis for planning a new campus. *Landscape Architecture.* 60(1):31−33.

Pound, C. E., and Egler, F. E. 1953. Brush control in southeastern New York: fifteen years of stable tree-less communities. *Ecology* 34(1):63−73.

Preston, F. W., and Norris, R. T. 1947. Nesting heights of breeding birds. *Ecology* 28:241−273.

Provost, M. W. 1948. Marsh blasting as a wildlife management technique. *J. Wildl. Mgmt.* 12(4):350−387.

Purves, D. 1966. Contamination of urban garden soils with copper and boron. *Nature (Lond.)* 210:1077.

———. 1967. Contamination of urban garden soils with copper, boron and lead. *Plant and Soil* 26(2):380−382.

Rackham, O. 1967. The history and effects of coppicing as a woodland practice. In *The Biotic Effects of Public Pressure on the Environment,* ed. E. Duffey. Monks Wood Exp. Sta. (England), Symp. 3:82−93.

Ragland, W. L., and Gorham, J. R. 1967. Tonsillar carcinoma in rural dogs. *Nature (Lond.)* 214(5091):925−926.

Rawley, K., and Peucker, T. K. 1970. Park awareness and park use in cities. In *The Geographer and Society,* eds. W. R. D. Sewell and H. D. Foster. Western Geogr. Series. Vol. I. pp. 125−133.

Recher, H. F. 1969. Bird species diversity and habitat diversity in Australia and North America. *Amer. Nat.* 103:75−80.

Reif, J. S., and Cohen, D. 1970. Canine pulmonary disease and urban environment. II. Retrospective radiographic analysis of pulmonary disease in rural and urban dogs. *Arch. Environ. Health* 20(6):684–689.

Reynolds, F. L., and Sankey, J. 1967. Public pressures on nature reserves and S.S.S.I's near large urban centres. In *The Biotic Effects of Public Pressures on the Environment*, ed. E. Duffey. Monks Wood Exp. Sta. (England), Symp. 3:162–168.

Ribaut, J. P. 1970. The need for green space. *Nature in Focus*, Spring, 1970: 2–13.

Rich, S. 1964. Ozone damage to plants. *Ann. Rev. of Phytopath.* 2:253–266.

Ridpath, M. G., Thearle, R. J. P., McCowan, D., and Jones, F. J. S. 1961. Experiments on the value of stupefying and lethal substances in the control of harmful birds. *Ann. Appl. Biol.* 49:77–101.

Risebrough, R., Florant, G. L., and Berger, D. D. 1970. Organochlorine pollutants in peregrines and merlins migrating through Wisconsin. *Can. Field Nat.* 84(3):247–253.

Robel, R. J., and Harper, W. R. 1965. Energy content and retention by ragweed and sunflower seeds during fall and winter. *Trans. Kans. Acad. Sci.* 68(3):401–405.

Rublowsky, J. 1967. *Nature in the City*. New York: Basic Books. 152 pp.

Ruhling, A., and Tyler, G. 1968. An ecological approach to the lead problem. *Bot. Notiser* 121:321–342.

Rumsey, R. L. 1970. Woodpecker nest failures in creosoted utility poles. *Auk* 87:367–369.

Safdie, M. 1967. Habitat '67. In *Environment for Man: The Next Fifty Years*, ed. W. Ewald. Bloomington, Indiana: Indiana U. Press. pp. 253–259.

Sage, B. L. 1970. The winter population of gulls in the London area. *Lond. Bird Rep.* 33:67–80.

Salisbury, E. J. 1945. The flora of bombed areas. *Nature (Lond.)* 151:462–466.

Salter, R. G. 1960. Trees and buildings in urban areas. *Quart. J. Forestry* 54(3):208–220.

Sanderson, R. 1968. The changing status of birds in Kensington Gardens. *Lond. Bird Rep.* 32:63–80.

Satchell, J. E. 1967. Lumbricidae. In *Soil Biology*, eds. A. Burges and F. Raw. London: Acad. Press. pp. 259–322.

Satterthwaite, A., and Marcou, G. T. 1969. Open space, recreation, and conservation. In *Principles and Practice of Urban Planning*, eds. W. I. Goodman and E. C. Freund, Washington, D.C.: Intl. City Managers Assoc. pp. 185–207.

Sawyer, E. J. 1969. Homes for wildlife, baths and feeding shelters; how to make them and where to place them. Cranbrook Inst. of Science. Bull. 1. 6th ed. Bloomfield Hills, Michigan. 36 pp.

Schoener, T. W. 1968. Sizes of feeding territories among birds. *Ecology* 49(1):123–141.

Scott, T. G. 1938. Wildlife mortality on Iowa highways. *Amer. Midl. Nat.* 20:527–539.

Shetron, S. G., and Duffek, R. 1970. Establishing vegetation on iron mine tailings. *J. Soil Water Conserv.* 25(6):227–230.

Shomon, J. J. 1971. *Open Land for Urban America*. Baltimore: The Johns Hopkins Press. 171 pp.

Shomon, J. J., Ashbaugh, B. L., and Tolman, C. D. 1966. *Wildlife habitat improvement*. New York: Natl. Audubon Soc., Nature Centers Div. 96 pp.

———. 1970. More wildlife for uban America. *The Conservationist* 24(4):2–7.

Shorten, M. 1962. Squirrels: their biology and control. Min. Agr. Fish. and Food. Bull. 184. HMSO, London. 44 pp.

Simms, E. 1962. A study of suburban bird-life at Dollis Hills. *Brit. Birds* 55(1):1–36.

Skye, E. 1958. Luftfororeningars inverkan pa busk-och bladlavfloran kring skifferoljeverket i märkes kvarntorp. *Svensk Bot. Tidsk.* 52:133–190.

———. 1968. Lichens and air pollution. Acta Phytogeogr. Sueccia 52. Uppsala. 23 pp.

Slatyer, R. O. 1970. Man's use of the environment–the need for ecological guidelines. *Aust. J. Sci.* 32(4):146–153.

Smith, E. R. 1970. Evaluation of a leveed Louisiana marsh. *Trans. N. Amer. Wildl. Conf.* 35:265–275.

Smith, N. C. (ed.) (n.d.) First professional level conference on open space preservation methods. Proc. of 1967 Lake Minnewaska New York Open Space Action Comm., New York. 67 pp.

Smith, W. H. 1970. Trees in the city, *J. Am. Inst. of Planners* 36:429–436.

Smith, L. P., and Wheeler, F. G. 1970. A nature conservation plan for an urban development; Estover Estate, Plymouth. Devon Trust for Nature Conservation. 33 pp.

Snow, D. W. 1958. *A Study of Blackbirds*. London: G. Allen & Unwin, 191 pp.

———. 1967. Population changes of some common birds in gardens. *Brit. Birds* 60(8):339–341.

Snow, D. W., and Mayer-Gross, H. 1967. Farmland as a nesting habitat. *Bird Study* 14(1):43–52.

Solmon, V. E. F. 1966. The ecological control of bird hazards to aircraft. *Proc. 3rd Bird Control Seminar.* Bowling Green State U. (Ohio). pp. 38–52.

Soulier, L. 1968. *Espaces verts et urbanisme.* Paris: Centre de Recherche d'Urbanisme, 275 pp.

Southern, H. N. (ed.) 1964. *The Handbook of British Mammals.* Oxford: Blackwell Sci. Publns. 465 pp.

Spencer, R. and Gill, D. 1972. Wildlife inventory of four selected Edmonton ravines; a progress report to Canadian Wildlife Service Western Region under Contract CWS 59 03002. Typescript. 32 pp.

Spilhaus, A. 1971. The next industrial revolution. Can. Impl. Bank of Commerce. Commercial Letter No. 1. 8 pp.

Spooner, C. S. Jr., and Yeager, L. E. 1942. Potential wildlife habitat on the Illinois prairie and some problems of restoration. *J. Wildl. Mgmt.* 6(1):44–54.

Staude, J. H. 1966. Bisherige ergebnisse eines versuches zur steigerung der siedlungsdichte hochlenbrustender vogelarten in reinen fichtenwaeldern im gebiet der Fuchskante, Westerwald. [Results of an experiment to increase breeding density of hole nesting birds in spruce forests.] *Emberiza* 1(3):79–92.

Stern, A. C. (ed.). 1968. *Air Pollution.* 2nd ed. Vols. I, II, III. New York: Academic Press.

Stockinger, H. E., and Coffin, D. L. 1968. Biologic effects of air pollutants. In *Air Pollution,* ed. A. C. Stern. New York: Academic Press. Vol. I, pp. 445–546.

Strawinski, S. 1963. Ptaki miasta Torunia. [The birds of the town of Torun.] *Acta Ornithol.* (Warsaw) 7(5):115–156. [Russian and Engl. Summ.]

Streeter, D. T. 1968. Countryside conservation–the role of county trusts. *Soc for Promotion of Nat. Reserves Hdbk. 51st Ann. Rep.* (British Museum [Nat. History]).

Strong, A. L. 1965. *Open Space for Urban America.* U.S. Govt. Printing Office. Washington, D.C. 154 pp.

Stuewer, F. W. 1943. Racoons: Their habits and management in Michigan. *Ecol. Monogr.* 13:203–257.

Sugden, L. G., and Benson, D. A. 1970. An evaluation of loafing rafts for attracting ducks. *J. Wildl. Mgmt.* 34(2):340–343.

Summers-Smith, D. 1963. *The House Sparrow.* London: Collins. 269 pp.

Syratt, W. J., and Wanstall, P. J. 1969. The effect of sulfur dioxide on epiphytic bryophytes. In *Air Pollution, 1969.* Wageningen. pp. 79–85.

Tabor, E. C. 1965. Pesticides in urban atmospheres. *J. Air. Polln. Control Assoc.* 15:415–418.

———. 1966. Contamination of urban air through the use of insecticides. *Trans. N. Y. Acad. Sci.* 28(5):569–578.

Tankel, S. B. 1963. The importance of open space in the urban pattern. In *Cities and Space: The Future Use of Urban Land,* ed. L. Wingo. Baltimore: The Johns Hopkins Press. pp. 57–71.

Tansy, M. F., and Roth, R. P. 1970. Pigeons: a new role in air pollution. *J. Air Polln. Control Assoc.* 20(5):307–309.

Tarrant, K. R., and Tatton, J. O'G. 1968. Organo-chlorine pesticides in rainwater in the British Isles. *Nature (Lond.)* 219:725–727.

Teagle, W. G. 1967. The fox in the London suburbs. *Lond. Nat.* 46:44–68.

———. 1969. The badger in the London area. *Lond. Nat.* 48:48–75.

Tebbens, B. D. 1968. Gaseous pollutants in the air. In *Air Pollution,* ed. A. C. Stern. New York: Academic Press. Vol. I. pp. 23–46.

Tendron, G. 1964. *Effects of pollution on animals and plants.* General Rep. of European Conf. on air pollution. Council of Europe. Strasbourg. pp. 25–70.

Terres, J. K. 1953. *Songbirds in Your Garden.* New York: Crowell. 274 pp.

Thearle, R. J. P. 1968. Urban bird problems. In *The Problems of Birds as Pests,* ed. R. K. Murton and E. N. Wright. London: Academic Press, pp. 181–197.

Thomas, M. D. 1965. The effects of air pollution on plants and animals. In *Ecology and the Industrial Society,* ed. G. T. Goodman *et al.* New York: J. Wiley and Sons, pp. 11–33.

Tischler, W. 1956. *Structure of landscape, protection of nature and plant protection.* Intern'l. Union Conserv. Nature and Nat. Res. 6th Tech. Meeting, Edinburgh. pp. 203–205.

Trent, C. 1965. *Greater London: Its Growth and Development through 2000 Years.* London: Phoenix House. 282 pp.

Tunnard, C., and Pushkarev, B. 1963. *Man-made America: Chaos or Control?* New Haven: Yale U. Press. 479 pp.

Turcek, F. J. 1960. on the damage by birds to power and communication lines. *Bird Study* 7(4):231–236.

Turril, W. B. 1948. *British Plant Life.* The New Naturalist Series. London: Collins. 315 pp.

Tyner, E. H., Smith, R. M., and Galpin, S. L. 1948. Reclamation of strip mined areas in W. Virginia. *J. Amer. Soc. Agron.* 40:313–323.

Uhler, F. M. 1956. New habitats for waterfowl. *Trans. N. Amer. Wildl. Conf.* 21:453–468.

U.S. Dept. Agric. 1969. Wildlife habitat improvement handbook. Forest Service Hdbk. 2609.11. Washington, D.C. 53 pp.

U.S. Dept. Interior. 1968. *Man and Nature in the City.* Symposium. Bureau Sports Fisheries and Wildlife. Washington, D.C. 1968. 91 pp.

U.S. Dept. Interior. (n.d.) Tinicum Marsh preservation: a proposal. Bureau Sports Fisheries and Wildlife. Washington, D.C. 12 pp.

Verts, B. J. 1957. The population and distribution of two species of *Peromyscus* on some Illinois strip-mined land. *J. Mammal.* 38(2):53–59.

Vesey-Fitzgerald, B. 1967. *Town Fox, Country Fox.* London: Andre Deutsch. 160 pp.

de Vos, A., and Petrides, G. A. 1967. Biological effects caused by terrestrial vertebrates introduced into non-native environments. Intern'l. Union Conserv. Nature and Nat. Res. 10th Tech. Meeting, Lucerne. 1966. pp. 113–119.

Vosburgh, J. 1966. Deathtraps in the (England) flyways. In *Birds in Our Lives,* ed. A. Stefferud. U.S. Dept. Interior. Washington, D.C. pp. 364–371.

———. 1968. Living with your land. Cranbrook Inst. of Science. Bull. 53. Bloomfield Hills, Michigan. 149 pp.

Wallace, G. J., Nickell, W. P., and Bernard, R. F. 1961. Bird mortality in the Dutch Elm disease program in Michigan. Cranbrook Inst. of Science. Bull. 41. Bloomfield Hills, Michigan. 44 pp.

Ward, R. K. 1964. Tolerance of weeds of waste areas to residual-type herbicides. *Proc. 17th New Zealand Weed and Pest Control Conf.* pp. 215–221.

Watt, A. S. 1955. Bracken versus heather: a study in plant sociology. *J. Ecology* 43:490–506.

Way, J. M. 1967. The influence of management on the flora of roadside verges. In *The Biotic Effects of Public Pressures on the Environment,* ed. E. Duffy. Monks Wood Exp. Sta. (England), Symp. 3:151–155.

Way, J. M. 1969. Toxicity and hazards to man, domestic animals and wildlife from some commonly used auxin herbicides. *Residue Reviews* 26:37–62.

———. 1970. Wildlife on the motorway. *New Sci.* 47:536–537.

Webb, L. J. 1968. Biological aspects of forest management. *Proc. Ecol. Soc. of Australia* 3:91—95.

"Webfoot" (pseudonym). 1871. *Fore and Aft; or Leaves from the Life of an Old Sailor.* Boston. 387 pp.

Weddle, A. E. 1967. *Techniques of Landscape Architecture.* New York: American Elsevier. 226 pp.

Weinberg, G. M. 1970. The nation as a park. *Nat. Parks and Consn. Mag.* 44(272):21.

Welch, R. C. 1968. Coppicing and its effect on woodland invertebrates. *J. Devon Trust for Nat. Conserv.* 13:969—973.

Wells, T. C. E. 1965. Chalk grassland nature reserves and their management problems. *Soc. for Promotion of Nat. Reserves Hdbk. 48th Ann. Rep.* (British Museum [Nat. Hist.]). pp. 62—70.

Westhoff, V. 1970. New criteria for nature reserves. *New Sci.* 46(697):108—113.

Weston, R. L., Gadgil, P. D., Salter, B. R., and Goodman, G. T. 1965. Problems of re-vegetation in the lower Swansea Valley, an area of extensive industrial dereliction. In *Ecology and the Industrial Society,* eds. G. T. Goodman *et al.* New York: J. Wiley and Sons. pp. 297—325.

Wheeler, A. 1969a. Fish-life and pollution in the lower Thames: A review and preliminary report. *Biol. Conserv.* 2(1):25—30.

Wheeler, F. G. 1969b. Nature conservation within the city boundary. *J. Devon Trust for Nat. Conserv.* 20:853—854.

Whyte, R. O. and Sisam, J. W. B. 1949. The establishment of vegetation on industrial waste land. Commonwealth Agr. Bur. Joint Publn. 14. Oxford. 78 pp.

Whyte, W. H. 1962. *Open space action.* Outdoor Recreat'n. Resources Rev. Comm'n. Rep. 15. Washington, D.C. 107 pp.

———. 1964. *Cluster Development.* Amer. Conserv'n. Assoc. New York. 130 pp.

———. 1968. *The Last Landscape.* New York: Doubleday and Co. 376 pp.

Wilber, C. G. 1969. *The Biological Aspects of Water Pollution.* Springfield, Illinois: C. C. Thomas. 298 pp.

The Wildfowl Trust. 1956. Some aspects of the management of wildfowl refuges in Britain. Intern'l. Union Conserv. Nature and Nat. Res. 6th Tech. Meeting, Edinburgh. pp. 65—71.

Will, G. C., and Crawford, G. I. 1970. Elevated and floating nest structures for Canada Geese. *J. Wildl. Mgmt.* 34(3):583—586.

Williams, K. 1964. Tawny owls feeding young on fish. *Brit. Birds* 57:202—203.

Williamson, K. 1967. Some aspects of the scientific interest and management of scrub on nature reserves. In *The Biotic Effects of Public Pressures on the Environment*, ed. E. Duffey. Monks Wood Exp. Sta. (England), Symp. 3:94‒100.

———. 1969. Bird communities in woodland habitats in Wester Ross, Scotland. *Quart. J. Forestry* 63(4):305‒328.

Wilmers, P. 1971. The good life in Virginia. *T. C. Planning* 39(1):73‒78.

Wingo, L., Jr. 1963. Urban space in a policy perspective: An introduction. In *Cities and Space*, ed. L. Wingo, Jr. Baltimore: The Johns Hopkins Press. pp. 3‒21.

Wolman, M. G., and Shick, A. P. 1967. Effects of construction on fluvial sediment, urban and suburban areas of Maryland. *Wat. Resources Res.* 3(2):451‒464.

Woodbury, C. 1966. The role of the regional planner in preserving habitats and scenic values. In *Future Environments of North America*, eds. F. F. Darling and J. P. Milton. New York: Natural History Press, pp. 568‒587.

Woolfenden, G. E., and Rohwer, S. A. 1969. Breeding birds in a Florida suburb. *Bull. Fla. State Mus. Biol. Sci.* 13(1):1‒83.

Woollacott, A. 1961‒7. Notes on the distribution and ecology of reptiles and amphibians in the Erewash Valley area of Nottinghamshire and Derbyshire. *Brit. J. Herpetol.* 3:83‒85.

de Worms, C. G. M. 1965. An analysis of the Macrolepidoptera recorded from the garden of Buckingham Palace. *Lond. Nat.* 44:77‒81.

Wright. E. N. 1967. Bird control on airfields. In *Proc. 2nd Brit. Pest Control Conf.* Indust. Pest Control Assoc. London. pp. 39‒41.

Yalden, D. W. 1965. Distribution of reptiles and amphibians in the London area. *Lond. Nat.* 44:57‒69.

Yalden, D. W., and Jones, R. 1970. The food of suburban tawny owls. *The Naturalist* 914:87‒90.

Yapp, W. B. 1966. Bring back the bear. *Wildlife and Countryside* 257:4.

Yeager, L. E. 1941. Wildlife management on coal-stripped land. *Trans. N. Amer. Wildl. Conf.* 5:348‒353.

Young, H. 1949. A comparative study of nesting birds in a 5-acre park. *Wilson Bull.* 61:36‒40.

Young, S. P. 1953. Coyotes. In *Wildlife Management*, ed. R. E. Trippensee. New York: McGraw-Hill Book Co. pp. 97‒102.

Young, S. P., and Jackson, H. T. 1951. *The Clever Coyote.* Washington, D.C.: The Wildl. Mgmt. Inst. 273 pp.

Zion, R. L. 1968. *Trees for Architecture and the Landscape.* New York: Reinhold Book Corp. 284 pp.

Zisman, S. B., Ward, D. B., and Powell, C. H. 1968. *Where Not to Build: A Guide for Open Space Planning.* Tech. Bull. 1. U.S. Dept. Interior, Bur. Land Mgmt. Washington, D.C. 160 pp.

Index

Composed in Press Roman by
The Composing Room

Printed offset by
Collins Lithography and Printing Co.
on 60 lb. WW&F Mystery Opaque Natural

Bound by Moore and Co. in Arrestox A 53500

DATE		